ADVENTURES IN CELESTIAL MECHANICS

ADVENTURES
IN CELESTIAL MECHANICS

A First Course in the Theory of Orbits

VICTOR G. SZEBEHELY

Richard B. Curran Centennial Professor of Engineering
University of Texas at Austin

 University of Texas Press, Austin

International Standard Book Number 0-292-75105-2
Library of Congress Catalog Card Number 88-51428

Copyright © 1989 by the University of Texas Press
All rights reserved
Printed in the United States of America

First Edition, 1989

Requests for permission to reproduce material from this
work should be sent to Permissions, University of Texas
Press, Box 7819, Austin, Texas 78713-7819.

*For reasons of economy and speed this volume has been
printed from camera-ready copy furnished by the author,
who assumes full responsibility for its contents.*

♾ The paper used in this publication meets the minimum
requirements of American National Standard for Information Sciences—Permanence of Paper for Printed Library
Materials, ANSI Z39.48-1984.

Dedication

This book is dedicated to the memory of Newton, who considered himself a boy playing on the seashore, and who has seen further by standing on the shoulders of giants. It was three hundred years ago in 1687 that Newton's *Philosophiae Naturalis Principia Mathematica* was published by the Royal Society of London. No more befitting dedication could possibly exist for this book than to the memory of Sir Isaac. His humility is reflected in this book by attempting to make it simple since, once the basic ideas are understood, even the most complicated problems can be presented in a simple manner and can be made enjoyable to those playing on the seashores and researching the frontiers of space. Newton also has seen further by standing on the shoulders of giants. This book intends to show the panorama by surveying our field of celestial mechanics from the shoulders of Newton, Euler, Lagrange, Poincaré, Whittaker, Einstein, and many others. Indeed, the view is unbelievably beautiful even if in some directions unsolved problems and inherent difficulties becloud the picture and limit the horizon. Instead of hiding these clouds, they are presented as challenges that should prompt us to search for the prettier shells.

Contents

Acknowledgments

It is easy to decide where to start the long list of acknowledgments. Sir Isaac's contribution to this book comes first since without the law of gravitation and his laws of motion it could not have been written. Albert Einstein's advice should be acknowledged immediately after Newton's contributions: Sitzfleisch, love, background and brain are the most important elements in accomplishments. (By the first item he meant perseverance, by the second dedication, by the third familiarity with the basic principles, and by the fourth mental ability.) My father showed me when I was five that equations are easier and simpler than multiplication tables and that it is easier and more satisfying sometimes to compute the proper trajectory of a football than to kick it. My thanks go to my students who never let me get away with hand waving, who asked questions and came up with new ideas. The contribution of time for research came from my endowed R. B. Curran Centennial Chair at the University of Texas and from my various grants from the National Science Foundation and other governmental organizations. The advice and detailed comments received from many of my colleagues, especially from Professor Raynor L. Duncombe, Dr. Jay H. Lieske, Dr. Joseph J.F. Liu, Professor John Prussing and from my research assistant Joseph Pojman were of considerable help.

The skillful typesetting and supervision of the typing was done by Mariou Barr using the *troff* typesetting package under the Unix operating system. The diligent typing was performed by Erica Garrison and by Rolando Nanez.

Introduction

The basic purpose of this book is to demonstrate the beauty of orbit mechanics and celestial mechanics and to share with the readers the pleasure of being in touch with the spirits of many of our heroes, from Aristotle to Brouwer. Our subject is not easy but enjoyably simple, once the basic ideas are understood. These fundamental ideas will be emphasized and will not be cluttered up with details, which are available in the immense literature of our field and which will be selectively pointed out with considerable care.

Following Einstein's advice we will develop perseverance by working on the analytical and numerical examples offered in every chapter. Emphasis on the basics, together with the historical remarks, will offer insight and background. The readers will bring the brains and the love of the subject, especially, when instead of feeling defeated and discouraged by unnecessary complications, they will enjoy following the simple footsteps of our giants.

Our subject is the study of the motions of natural and artificial bodies in space. This field is sometimes called celestial mechanics, which emphasizes the motion of natural bodies and is part of dynamical astronomy. Another name of our field is astrodynamics, which is mostly concerned with artificial satellites and space probes, and sometimes it is part of the engineering curriculum. Also, we often hear the terminology "orbit mechanics," which is rather descriptive, and it might include the astronomical and the engineering aspects.

The common and basic underlying ideas connecting the various aspects of our field are Newton's laws of dynamics and his law of gravity.

The first chapter discusses Earth satellites on circular orbits, and it shows that with relatively simple approaches important practical problems can be solved. The purpose of this chapter is to offer self-confidence to the reader and to demonstrate that elaborate and involved mathematics is not always necessary in orbit mechanics. This self-confidence will be supported with some examples, solutions of which will hopefully capture the reader. The second chapter offers a short historical review to show that the reader is indeed in a good company when he studies celestial mechanics. The third chapter discusses the basic problem of two gravitationally interacting bodies and shows applications for space probes and for members of the solar system. Chapter 4 treats elliptic orbits and shows the power of selecting the proper variables in dynamical problems. Not only will readers become experts in orbit changes, transfer orbits and other fundamentals in orbit mechanics, but they will learn some very impressive but basically simple concepts and terminologies inherited from classical celestial mechanics in Chapters 5 and 6. Hyperbolic and parabolic orbits are discussed in Chapter 7 offering some details concerning these high-speed orbits of interest in the dynamics of comets and in modern orbit mechanics. In Chapter 8 the classical theorem of Lambert is treated for elliptic, hyperbolic and parabolic orbits, showing, once again, how classical results can be used to solve problems of modern space dynamics. Chapter 9 introduces orbital elements and explains the reasons for their popularity in celestial mechanics. Chapter 10 discusses perturbation methods in an introductory level with the purpose of motivating the reader to dedicate his career to solving problems which in some respect are unsolvable. Chapter 11 offers the treatment of artificial satellites showing how the orbital elements change because of perturbations.

An interesting progress might be pointed out from Chapter 1 to Chapter 11. The approximate results concerning circular orbits of Earth satellites given in Chapter 1 are generalized and modified in Chapter 11 to include the effects of perturbations due to the non-spheriodicity of the Earth and to atmospheric drag.

The celebrated problem of three bodies is the subject of the last chapter, where we show that the so-called restricted form of this problem represents the basis for establishing lunar and interplanetary trajectories. The chapter of concluding remarks offers readers some true evaluations of what they have learned and what we still cannot do. Our pride will be balanced by humility when the limitations of our fields related to uncertainties and to non-integrabilities are revealed.

The text of the chapters is concluded with numerical and analytical examples discussed in detail and with unsolved problems to challenge the reader. The appropriate references are mentioned at the end of the chapters as well as in the Appendix.

The Appendix begins with the list of frequently used terms and their definitions. This is followed by the presently available "best" values of often used constants. Attention is directed to the uncertainties of these "constants" which might change as new observational results become available. An annotated list of major reference books is included in the Appendix to help us to consumate the love affair with our subject. The extensive Subject Index was prepared with considerable care to assist the reader to locate even those subjects which are mentioned but not treated in detail. The Name Index will help the reader to extend his knowledge at will and to become familiar with the pertinent literature.

The book opens with optimism and enthusiasm and closes with the challenges of presently unsolved problems and with the recognition of our limitations.

The following personal remark might serve one of the major purposes of this book, which is to help the reader to become infatuated with the subject of celestial mechanics. It was during the fall of 1943 that my Chairman at the Technical University of Budapest called me into his office and told me that the subject of my dissertation was the problem of three bodies. When I asked him what, specifically, I should be doing he looked at me and said, "Go away, and solve it. Read Whittaker's book!"

The reader should be filled in with a few details at this point in order to comprehend the story. In those days in most of the European universities the Chairman of the department was the absolute ruler and king, in charge of the assistants, adjuncts, graduate students, etc. He decided who qualified as a PhD student without conducting a formal qualifying examination. (In fact this was the case also in the 1960s at Yale University where Dirk Brouwer, as the Chairman of the Department of Astronomy decided which graduate students qualified and which were sent away.) The next information the reader needs to realize is that the assistant in these days did not ever argue with the professor but did as he was told or just left the University.

So when I was given my assignment, I was first very happy that I "qualified" and ran to the library to read Whittaker's book. Apparently, my king did not realize that my English was not adequate to read this book, but, of course, I could not disturb him with such minor details and started working on Whittaker's *Analytical Dynamics*. The chapters on the problem of three bodies were absolutely fascinating, and I could not put down the book until I read it several times with a pencil in my hand to figure out the missing details. I believe this was the point in my life when I fell in love with celestial mechanics. I started playing less soccer, cut back on movies, even reduced time chasing girls and started thinking on what I learned from Whittaker.

The most important thing I immediately realized was Poincaré's non-integrability principle. As the reader will see later on, this principle means that no generally valid solution can be found for the problem of three bodies. So I went back to my ruler and after reminding him who I was and what my assignment was, told him that the problem cannot be solved and I even referred to Poincaré, partly showing off and partly hoping that he would not execute me on the spot. Well, he actually smiled, which I had never seen before, and told me to go and solve the problem of three bodies. I was standing in front of his desk and could not move. I realized that my love of dynamics was not reciprocated and that my infatuation led me to a catastrophy since I would never receive my PhD degree. My king must have noticed my silent desperation and said: " I told you to solve the problem of three bodies. I never told you to find the general solution or to solve the gravitational problem of three bodies. Read Radau's article in Comptes Rendus."

So I ran to the library and immediately realized that this time I was supposed to read French, which I did with considerable difficulties. I also established that the variables invented by Radau allowed us to find several new particular solutions of the problem of three bodies when the forces acting between the bodies were not inversely proportional to the squares of the distances (as in the case of gravity) but directly proportional to the cube of the distances. My dissertation was finished in March 1946 and a version was published in 1952 in the U.S.

My love affair was reinforced and I learned English and French. The most important thing I learned was that in science careful attention must be given to the formulation of problems and that generalizations sometimes offered new results. I am still looking for the physical application of this strange force law and I am convinced that not finding it is my limitation and not nature's.

ADVENTURES IN CELESTIAL MECHANICS

Chapter 1. Circular Orbits of Earth Satellites

A simple approximate method to study circular orbits of satellites around the Earth is described. The purpose of this chapter is to show that orbit computations are not necessarily difficult or complicated and that the application of some basic principles of mechanics bears useful and immediately applicable results.

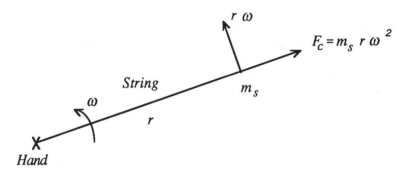

Figure 1.1. Centrifugal Force Balanced by String.

Figure 1.1 shows a body with mass m_s attached to the end of a string and rotated with an angular velocity ω. The centrifugal force acting on the body is proportional to the mass of the body, to the distance from the center of rotation and to the square of the angular velocity

$$F_c = m_s r \omega^2 .$$

The centrifugal force is balanced by the string, or by the hand which is holding the other end of the string. The angular velocity is ω, consequently, the velocity of the body moving on a circular orbit is $r\omega$, where r is the distance from the center of revolution to the body. The body will not leave the circular orbit unless we either stop the rotation or we let go of the string. We shall now apply this simple model to a satellite circling the Earth at a certain elevation. The force which is holding the satellite on its orbit is gravity which replaces the string in the previous example. Newton's law of gravity gives the force which keeps the satellite on its orbit. This force (F_g) is proportional to the gravitational constant (G), to the mass of the Earth (m_E), to the mass of the satellite (m_s) and inversely proportional to the square of the distance between the center of the Earth and the satellite (r). Equation (1.1) gives this relation as:

$$G \frac{m_E m_s}{r^2} = F_g \ . \tag{1.1}$$

This force (F_g) is now equated to the previously found centrifugal force, F_c giving

$$m_s r \omega^2 = G \frac{m_E m_s}{r^2} \ . \tag{1.2}$$

This is the basic equation of circular motion of a satellite. Note that in this equation the mass of the satellite may be cancelled. Consequently, if one is interested in finding the mass of a body (such as the Earth), one must use that body as the central body and not the body whose motion is being observed. This idea will be discussed in more detail later on and it should be remembered that the motion of planetary satellites (artificial or natural) are of considerable importance when the masses and the mass distributions of their planets are to be determined. From Eqn. (1.2) we have

$$r \omega^2 = \frac{G m_E}{r^2} \ . \tag{1.3}$$

At this point we replace ω as the symbol for the angular velocity by n , which is called the mean motion in orbit mechanics. (The use of the special and proper terminology separates the experts from the laymen.) The mean motion becomes

$$n = \left[\frac{G m_E}{r^3} \right]^{\frac{1}{2}} \ . \tag{1.4}$$

The circular velocity can also be obtained from this equation by multiplying both sides by r ,

$$v_c = nr = \left[\frac{Gm_E}{r} \right]^{\frac{1}{2}} . \tag{1.5}$$

The velocity around the Earth or the mean motion allows the computation of the period of a satellite. The length of the path traveled during one revolution is $2\pi r$ and since the satellite is moving with velocity v_c it will take P seconds to go around once, or

$$2\pi r = P v_c . \tag{1.6}$$

Here P is the period,

$$P = \frac{2\pi r}{v_c} . \tag{1.7}$$

Note that the period can be also obtained from the mean motion by writing

$$P = \frac{2\pi}{n} . \tag{1.8}$$

The circular velocity of a satellite can be computed from the mass of the body around which the satellite is in orbit, from the distance to the center of rotation and from the constant of gravity.

Equation (1.5) might be slightly modified by realizing that usually the altitude of a satellite above the surface of the Earth is given, rather than its distance from the center of the Earth. This relation is

$$r = R_E + h , \tag{1.9}$$

where R_E is the radius of the Earth and h is the altitude of the satellite above the Earth. So Eqn. (1.5) becomes

$$v_c = \left[\frac{Gm_E}{R_E + h} \right]^{\frac{1}{2}} . \tag{1.10}$$

High altitude satellites move slowly since the centrifugal force to be compensated by the gravitational attraction of the Earth is small. A very special satellite can be established which when launched in the plane of the Earth's equator will move with the Earth. Its altitude should be such that its velocity will keep up with the rotation of the Earth. The problem of finding the right altitude is relegated to the examples at the end of this chapter. These satellites are known as geosynchronous or stationary satellites.

Squaring both sides of Eqn. (1.4) and multiplying by r^3 we get

$$n^2 r^3 = G m_E . \tag{1.11}$$

This is a very famous equation in celestial mechanics and it is known as a form of Kepler's law. Eqn. (1.11) is a very good approximation to the exact relation. The approximate result assumes that the mass of the satellite can be neglected when compared to the mass of the central body. The derivation of the exact relation for circular motion utilizes Fig. 1.2.

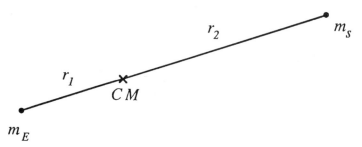

Figure 1.2. Motion of a Satellite around the Center of Mass of the System.

The satellite and the Earth are moving around the center of mass of the Earth-satellite system. Since the mass of the Earth is always many orders of magnitude larger than the mass of the satellite, the center of mass of the system is at the center of the Earth for all practical purposes. As another interesting example consider a binary star or a binary asteroid where two stars or two asteroids with comparable masses are revolving around each other. The distances from the center of mass are r_1 and r_2, the masses are m_1 and m_2. In our original problem $m_1 = m_E$ and $m_2 = m_s$. The forces acting on the satellite are balanced if

$$r_2 n^2 m_s = G \frac{m_E m_s}{r^2} . \tag{1.12}$$

The corresponding equation for the Earth is

$$r_1 n^2 m_E = G \frac{m_E m_s}{r^2} . \tag{1.13}$$

In the first equation m_s and in the second equation m_E are cancelled. Since the center of mass is fixed in the system, we have

$$r_2 m_s = r_1 m_E . \tag{1.14}$$

Adding the two previous equations (1.12 and 1.13) we have

$$(r_1 + r_2)n^2 = G\,\frac{m_E + m_s}{r^2}\,, \qquad (1.15)$$

where r is $r_1 + r_2$ since this is the distance between the attracting bodies. Eqn. (1.15) may be written as

$$n^2 r^3 = G(m_E + m_s)\,. \qquad (1.16)$$

This is the exact form of Kepler's law for circular orbits and we see that the previously obtained Eqn. (1.11) needs a modification, whereby instead of the mass of the Earth we have the sum of the mass of the Earth and the mass of the satellite on the right-hand side. Clearly, in satellite dynamics this makes no difference and consequently Eqn. (1.11) is correct. On the other hand if we study bodies revolving around each other with comparable masses Eqn. (1.16) should be used.

To evaluate the approximation, Kepler's law may be written as

$$n^2 r^3 = Gm_E \left[1 + \frac{m_s}{m_E} \right]. \qquad (1.17)$$

If the mass of the satellite is 1 ton, the m_s/m_E ratio becomes of the order of 10^{-22}, which is the error made when the m_s/m_E term is neglected.

It might be surprising to find that a number of practically important problems can be treated using the simple results of Chapter 1. Some of these are discussed in the following examples.

Introductory chapters from Danby's (1962), McCuskey's (1963) and Ryabov's (1959) books are recommended. (See references listed in the Appendix.)

EXAMPLES

1. Consider a satellite on a circular orbit at an altitude of 100 kilometers (= 100,000 meters = 328,100 ft = 62.137 statute miles). For the computation of the circular velocity of the satellite the values of the constants are given in the Appendix. Eqn. (1.10) gives 7.844 kilometers/second or 17,548 miles/hour for the circular velocity. The satellite is moving approximately 300 times faster than the allowable speed limit for cars. In this result no drag effects due to the Earth's atmosphere are included and the Earth is assumed to be a homogeneous spherical body. The corrections caused by these effects are small for a few revolutions and will be discussed later. The mean motion using $n = v_c/r$ becomes $1.211 \times 10^{-3}\ radians/sec$ = 4.163

degrees/min, and the period is $P = 2\pi/n = 5188.43 \ sec =$ 1 *hour* 26 *min* 28.4 *sec*.

If the altitude is 500 km, the above results change little. In fact the circular velocity becomes $v_c = 7.613 \ km/sec$, the mean motion is $n = 1.107 \times 10^{-3} \ rad/sec$ and the period becomes $P = 1 \ hour \ 34 \ min \ 34 \ sec$.

2. The second example treats the problem of a 24-hour (geosynchronous or geostationary) satellite. This satellite has a period of 24 hours; consequently, if its orbital plane is in the Earth's equatorial plane, then it will revolve around the Earth so that it will be always above the same point of the Earth. This satellite is used for communication purposes. From the above period the mean motion can be computed. From the mean motion the circular velocity as well as the height or the altitude of the satellite can be obtained. Since the period is $P = 24 \ hr = 86400 \ sec$, the mean motion becomes

$$n = \frac{2\pi}{P} = 7.272205 \times 10^{-5} \ rad/sec \ .$$

The period and the altitude are connected by

$$P = \frac{2\pi}{\sqrt{Gm_E}} \ \sqrt{(R_E + h)^3} \ ,$$

from which the altitude becomes

$$h = \left[\frac{P}{2\pi} \right]^{3/2} (Gm_E)^{1/3} - R_E = 35862 \ km \ ,$$

or 22275 miles. Equation (1.18) follows from Eqns. (1.7), (1.9) and (1.10).

Another approach is to use the relation between the mean motion and the altitude:

$$(R_E + h)^3 n^2 = Gm_E \ ,$$

from which the altitude may be computed as

$$h = \left[\frac{Gm_E}{n^2} \right]^{1/3} - R_E \ .$$

The circular velocity of the satellite is given by $v_c = n(h + R_E)$. This gives $3.072 \ km/sec = 11058 \ km/h$ or $6871 \ miles/h$.

The triangles of Fig. 1.3 show the relation between the circular velocity of the satellite which will now be denoted by v_{cs} and the circular velocity of a point on the equator, v_{cE} as

$$\frac{v_{cs}}{v_{cE}} = \frac{h + R_E}{R_E} = \frac{h}{R_E} + 1 .$$

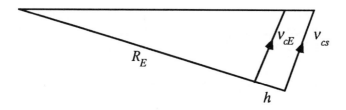

Figure 1.3. Geosynchronous Satellite.

The circular velocity of the equator is obtained from

$$v_{cE} = 2\pi \frac{R_E}{P_E} ,$$

where $P_E = 24 \ hour$. This gives $v_{cE} = 1668 \ km/h$ or $1038 \ miles/h$. This allows us an alternate way to compute the circular velocity of a synchronous satellite using the previously obtained value for the altitude:

$$v_{cs} = \left[\frac{h}{R_E} + 1 \right] v_{cE} .$$

3. The third example is related to the approximately circular lunar orbit. The period is approximately 27.32 days from which the mean motion, the altitude above the Earth, and the velocity might be computed.

The mean motion is given by

$$n = \frac{2\pi}{P} \ ,$$

where P is the lunar period. This gives $n = 2.66 \times 10^{-6} \ rad/sec =$ 13.18 $degrees/day$.

The distance between the center of the Earth and the center of the moon (r) is related to the mean motion by

$$r^3 n^2 = G(m_E + m_M) \ ,$$

which gives $r = 384400 \ km$.

The circular velocity of the moon may be obtained from $v_c = nr$ which gives 3594 km/h or 2233 miles/h. An easy to remember approximate value is 1 km/sec.

4.   The fourth exercise will be the preparation of a plot showing the periods (P in hours) of satellites in circular orbits around the Earth versus their altitude (h) up to 1000 km. The basic relation is given by Eqn. (1.18).

When the constants are substituted we have

$$P = 1.4082 \left[1 + \frac{h \, (km)}{6378.12} \right]^{3/2} .$$

Fig. 1.4 shows the plot of the $P(h)$ relation. This is approximately a straight line which suggests the next example.

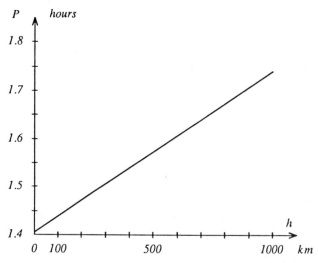

Figure 1.4. Period Versus Altitute for Earth Satellites.

5. This example will show the derivation of a simple, approximately linear relation between the period and the altitude for Earth satellites in circular orbits. Note that the exact relation is given by Eqn. (1.18) from which we may obtain

$$P = \frac{2\pi}{\sqrt{Gm_E}} R_E^{3/2} \left[1 + \frac{h}{R_E} \right]^{3/2} .$$

The factor in front of the parenthesis represents the period a satellite with zero altitude would have, $P_o = 1.40815 \; hours$. The equation for the period might be written as

$$P = P_o \left[1 + \frac{h}{R_E} \right]^{3/2} .$$

The binomial expansion of the factor of P_o is

$$\left[1 + \frac{h}{R_E} \right]^{3/2} = 1 + \frac{3}{2} \frac{h}{R_E} + \frac{3}{8} \left[\frac{h}{R_E} \right]^2 \pm \cdots$$

Therefore, the linear relation between altitude and period becomes

$$P_L = P_o \left[1 + \frac{3}{2} \frac{h}{R_E} \right] .$$

The error between the exact and the approximate equations is $P - P_L$ and the relative error is

$$\frac{P - P_L}{P} .$$

If this error is to be less than 1 percent, we have

$$\frac{P - P_L}{P} \leq 0.01 .$$

In this equation P and P_L depend on the ratio h/R_E which we shall denote by x. In this way we have

$$\frac{(1+x)^{3/2} - (1 + \frac{3}{2}x)}{(1+x)^{3/2}} \leq 0.01 .$$

This inequality leads to a cubic equation for x which may be written as

$$x^3 + 0.704316x^2 - 0.060912x - 0.020304 \leq 0.$$

The only positive root is $x_o = 0.1887$ which means that if $x = h/R_E$ is less than the above value given for x_o, the error of using the linear equation instead of the exact equation for the period will be less than 1 percent. The range of applicability of the linear solution is from $h = 0$ to $h = 0.1887R_E$ or

$$0 \leq h \leq 1203.56 \, km.$$

Note that this problem can be solved by either iteration of the inequality

$$0.99 \, (1+x)^{3/2} \leq 1 + \frac{3}{2}x$$

or by using the well-known formula for the solution of cubic equations.

6. Newton computed the ratio of the Earth's gravitational force and the centrifugal force at the equator and obtained "approximately 300." This example is aimed at computing this ratio with the constants given in the Appendix.

   The centrifugal force (per unit mass) is

$$F_c = R_E \omega^2 \,,$$

where R_E is the Earth's equatorial radius and ω is the angular velocity of the Earth's rotation.

   The gravitational force per unit mass is

$$F_g = \frac{GM_E}{R_E^2} \,.$$

The ratio is

$$\frac{F_g}{F_c} = \frac{GM_E}{R_E^3 \omega^2} = 288.3 \,.$$

7. To compute the mean density of the planets and of the Sun the values given for their radii and masses in the Appendix can be used. Substituting into the formula for density

$$\rho = \frac{Mass}{Volume} = \frac{3M}{4\pi R_E{}^3}$$

and remembering that the density of water is 1 g/cm^3 the values computed will show the ratio of the density of the planets to that of water. The mean density of the Sun is 1.41 g/cm^3, and of the Moon is 3.34 g/cm^3. For the planets the values are given in the table below.

Planet	Mercury	Venus	Earth	Mars	Jupiter	Saturn	Uranus	Neptune	Pluto
density (g/cm^3)	5.13	4.97	5.52	3.94	1.33	0.69	1.56	2.27	4 ?
velocity (km/sec)	47.8	35	29.8	24.2	13.1	9.7	6.8	5.4	4.7?

8. The orbital velocities of the planets are computed from the formula

$$v = \left[\frac{G(m_S + m_P)}{a_P} \right]^{\frac{1}{2}},$$

where m_S and m_P are the masses of the Sun and of the planet and a_P is the mean distance of the planet. The values are given in the table above.

9. Compare the compactness of the solar system with that of the hydrogen atom. The compactness or denseness can be measured by the ratio of the distance (a) of a planet (electron) from the Sun (nucleus) to the radius R of the Sun (nucleus). For the hydrogen atom we have $a/R = 10^5$ and for the Sun-Earth combination we have $a/R = 200$ which value for the Sun-Jupiter combination becomes 1000. The corresponding volume ratio is $(a/R)^3$. This becomes 10^{15} for the atom and 10^9 for the Sun-Jupiter combination. We can conclude, therefore, that the solar system is more condensed than the atomic world by factor of 10^6.

Comparing revolutions, the hydrogen atom in ground state performs 10^{16} revolutions per second, corresponding to 3×10^{23} rev/year, while the Earth makes 1 rev/year and the corresponding value for Jupiter is 0.0843 rev/year.

So the 10^6 times denser solar system moves much slower, in fact by factor of 3.57×10^{24}, considering the number of revolutions performed by the two systems during the same time. Because of the popularity of supercolliders, i.e. accelerators, their corresponding 6×10^{10} revolution per year might also be mentioned. Considering the age of the solar system (about 10^9 years), the corresponding running time for an accelerator is 6 days. Conclusions are left to the reader.

PROBLEMS

1. Find the mean motion, altitude and circular velocity of a geosynchronous satellite using the precise period of the sidereal day: 23 hours, 56 minutes, 4.1 seconds.

2. Derive linear formulas showing the analytical dependence of the deviations in n, h and v_{cs} on a small change in the period: $\Delta n = f_1(\Delta T)$, $\Delta h = f_2(\Delta T)$ and $\Delta v_{cs} = f_3(\Delta T)$.

3. Compare the actual values of the semi-major axes of the planets as given in the "Table of Constants of the Planets" in the Appendix with the values computed from Bode's law (actually announced by J.D. Titius in 1766) according to which

$$a_n = (4 + 3 \times 2^n) \times 0.1 .$$

Here, a_n is the semi-major axis of the $n-th$ planet in astronomical units. Note that $n = -\infty, 0, 1, 2, 3,...$ and the value $n = 3$ corresponds to the orbits of minor planets such as Ceres, discovered in 1801 with the help of the "Titius-Bode law."

Chapter 2. Historical Review - The Company We Keep

The history of celestial mechanics is often divided in four parts. The first one representing about 2000 years begins with Aristotle and includes Ptolemaeus, Copernicus, Brahe, Galileo and Kepler.

The second part is probably the most significant from a scientific point of view, and it is often considered the classical period. It begins with Newton and includes Descartes, Leibnitz, Halley, Euler, Clairaut, D'Alembert, Lagrange, Laplace, Legendre, Gauss, Poisson, Encke and Hamilton.

The third part represents the "modern," 19th century contributors: Hill, Tisserand, Poincaré, Moulton, Whittaker and Birkhoff.

The fourth group is associated with the latest, 20th century contributors: Arnold, Brouwer, Duboshin, Herget, Herrick, Kolmogorov, Moser, Siegel, Wintner.

These lists are, by all means, not complete. A more systematic and compact list is given in a tabulated form at the end of this chapter. The conditions for inclusion in this list were that the contributors offered some generally accepted and used methods in celestial mechanics, that they were born before 1920, and died before 1988.

A few remarks are offered concerning the life of some of the participants to show how their work fits in the development of our field. Their major scientific contributions are discussed in more detail in the appropriate chapters of the text, and in the Appendix, where an annotated list of references is offered.

Aristotle (384-322 B.C.) was one of the great Greek philosophers introducing the idea of a geocentric solar system, in fact promoting the concept of a geocentric universe. Another of his "laws" was that bodies of different weights fell at different speeds. You can visualize the great man Aristotle walking around carrying a small stone in his right hand and a larger stone in his left hand. If he would have dropped these two bodies and measured their velocities or the time which it took for the bodies to hit the ground, he would have found that their velocities were equal. But this would have been science, and this would have been a scientific experiment in physics; so Aristotle did not drop the two bodies. On the other hand we should mention the name of Claudius Ptolemaeus (151-127 A.D.) who was one of the "scientific" proponents of the geocentric system for the universe, using epicycles to describe planetary motion.

Note that some other Greek philosophers in the 3rd century B.C. already suggested that the Sun might be the "center of the universe" instead of the Earth occupying this position. We might wish to call this the first appearance of the heliocentric hypothesis.

Nicolaus Copernicus (1473-1543) was born in Torun, Poland. His original name was Mikolaj Kopernik. He was probably the first scientific proponent of the heliocentric view, emphasizing the simplicity of this model. He was a student in Krakow (1491), in Bologna (1497) and at the University of Padua (1501). He settled as a canon of Frauenburg where his famous book *De Revolutionibus Orbium Coelestium* was written. He was afraid of ridicule and allowed publication of his book only after his death. The way Copernicus and his students proposed the heliocentric idea was that they suggested this as a convenient way of computing planetary orbits rather than as a physical fact. Nevertheless, the Catholic Church still put Copernicus' work on the index of prohibited books in 1616. Note that Copernicus considered all planets on circular orbits around the Sun.

The next significant steps were made by the German astronomer **Johannes Kepler** (1571-1630) who in addition to a heliocentric view introduced the idea of elliptic orbits of the planets. His three laws of planetary motion and the equation named after him, which connects the eccentric anomaly with the mean anomaly, are still in use today. Many of his results were based on the observations of the Dutch astronomer Tycho Brahe (1545-1601) whom he followed as court mathematician in Prague after his professorship in Graz was terminated in 1600.

Kepler encountered the well-known problem of scientists, which still exists today, when his sponsor, the Emperor Rudolf II, requested that he should prepare his horoscope. Kepler considered astrology as a source of income rather than science and became one of the first "principal investigators" satisfying the requests of a sponsor against his own better judgement offering "practical and applicable results."

Galileo Galilei (1564-1642) might be considered one of the fathers of dynamics and of the heliocentric view of the solar system. He was born in Pisa and became a lecturer of mathematics at the University of Pisa in 1589. From there he moved to Padua where he became professor of mathematics in 1592. His famous *Dialogo* was written in Florence and was published in 1632. His equally important *Discorsi* was published in 1638 in which he summarized his accomplishments in mechanics. He died in Arcetri, near Florence, where he spent the last eight years of his life in house arrest as punishment for his conflicts with the Church. A forerunner of Newton, Galileo offered verbal statements of some of Newton's laws and introduced many concepts of kinematics and dynamics. As an astronomer, he was one of the principal contributors of the discovery of the telescope which was used for observations supporting the heliocentric theory. His often quoted free fall experiments from the leaning tower of Pisa cannot be substantiated, but his observations concerning the period of oscillations of the chandelier in the cathedral appear in his written material and allowed him to propose the use of pendulums to regulate clocks. Today, we realize that because of the small amplitude of the oscillations, the observed period was independent of the amplitude and Galileo did not have to introduce nonlinear corrections.

He expressed his strong feelings of the importance of mathematical rationalism by stating that the "book of nature is written in mathematical characters."

Sir Isaac Newton's (1642-1727) contributions for many historians of science represent the beginning of a completely new dynamics and celestial mechanics, indeed, a new meaning of "science." This is most adequately summarized by the quotation, "Nature and nature's laws lay hid in night. God said 'let Newton be,' and all was light" by Alexander Pope.

Newton was born on Christmas day in 1642. He received his B.A. degree in Cambridge in 1665. His professor Isaac Barrow resigned, requesting that Newton should get his professorship in 1669. Newton's complete dedication to his work resulted in headaches, sleepless nights, irregular eating habits and finally in a nervous fatigue at 50 years of age. When working on the computations of the motion of the moon he actually mentions these problems in his notes.

Before he switched from science to administrative activities and became the Warden and in 1699 the Master of the Mint, his book, entitled *Philosophiae Naturalis Principia Mathematica*, was published in 1687 by the Royal Society of London. It is interesting to see how complicated dynamical problems can become at Newton's insistence on using geometry instead of calculus to solve them. This makes the *Principia* a very hard book to read and leads to the question of why Newton did not use calculus when he was one of its inventors? The answer might be that this was just one of the well-known cases, when publications do not exactly represent the process by which the results are arrived. Newton apparently formulated and solved some of these problems by calculus, but being afraid of criticism he described his work using geometry.

Newton's conflicts with Leibnitz concerning the discovery of calculus are well represented in the literature, and this might have been another reason why geometry dominates the *Principia*. Their controversy regarding the deterministic nature of dynamics and celestial mechanics is less known. Today Newtonian mechanics is sometimes erroneously associated with complete predictability in dynamics, which was Leibnitz' dogma and was not accepted by Newton. Laplace's demon enters the picture at this point knowing all initial conditions and all laws of nature and predicting the future. He takes the side of Leibnitz. (See note at the end of Chapter 2.)

In 1665 Newton left Cambridge because of the plague and went back to his birthplace where he could dedicate his time to undisturbed thinking. The unverified apple incident, which could have happened here, describes the importance of connecting seemingly unrelated phenomena; in this case falling stones (or apples) on one side and planetary motion on the other. In fact, Newton describes the idea leading to artificial satellites with the following thought experiment. If stones are thrown from the top of a mountain with small horizontal velocities, they will hit the ground, but as the velocity is increased, circular and elliptic orbits are obtained around the earth.

The conditions of creativity, concentration and peace were present in his yard. In addition, he succeeded in relating seemingly unrelated phenomena, arriving at the general theory of gravitation.

Newton became the president of the Royal Society at the age of 60 and was knighted by Queen Anne in 1705. He died in 1727 and is buried in the Westminster Abbey in London.

Since in this book we wish to concentrate on dynamics and celestial mechanics, for the description of Newton's many other significant scientific contributions (for instance his *Opticks* published in 1704), the reader is referred to the literature.

His laws of dynamics and his law of gravitation will be described here, since a few general historical comments might be appropriate.

His three laws of motion, forming the basis of dynamics, are:

1. Every body perseveres in its state of rest or uniform straight-line motion, unless it is compelled by some impressed force to change that state.

2. The change of motion is proportional to the motive force impressed and takes place in the same direction as the force.

3. Action is always contrary and equal to reaction.

There are many variations of these laws, some of them by Newton himself making changes and corrections. Several different presentations exist in the literature as the laws were translated from the original Latin text.

Once again the soundest language, mathematics, comes to our help. Using the concept of linear momentum (which Newton called motion) we can express the first and second laws by the equation

$$\frac{d(m\bar{v})}{dt} = \bar{F}.$$

Note that Newton did not mention acceleration when giving his laws of motion. For a constant value of the mass, the above equation becomes $m\bar{v} = \bar{F}$. Our textbooks use the concept of acceleration and give Newton's law as $m\bar{a} = \bar{F}$. This is of less generality than Newton's original formulation which is applicable to variable mass and, therefore, for rocket propulsion.

Newton's law of gravitation, as discussed in his *Principia* published in 1687, was mentioned before. The gravitational force acting between two bodies is proportional to the product of the masses and it is inversely proportional to the square of the distance between them. In vectorial form

$$\bar{F} = G\frac{m_1 m_2}{|\bar{r}|^3}\bar{r}.$$

Probably nothing describes Newton better than one of his own statements. "I seem to have been only like a boy playing on the seashore and diverting myself in now and then finding a smoother pebble or a prettier shell than ordinary while the great ocean of truth lay all undiscovered before me." Another quotation reveals even more humbleness: "If I have seen further, it is by standing on ye shoulders of giants." In order to balance his humility we also must quote Edmund Halley: "Nec fas est propius mortali attingere divos." ("Nearer the gods no mortal may approach.")

Edmund Halley (1656-1742), English astronomer, observed and calculated the orbit of the comet named after him using Newton's laws of motion. Studying several cometary orbits, he established the facts that, contrary to planetary orbits, some comets had large angles of inclination and that some had periodic orbits. Halley's contributions were numerous and important to celestial mechanics, but his insistence on and support of the publication of Newton's *Principia* represent probably his greatest influence on today's celestial mechanics.

Leonhard Euler (1707-1783) was born in Basel, Switzerland. He was a student of Johann Bernoulli (father of Daniel and Nicolas Bernoulli) and in 1727 went to St. Petersburg (now Leningrad) for 14 years and was associated with the Academy. From here, at the invitation of Frederick the Great, he went to Berlin for 25 years and returned to St. Petersburg at the invitation of the czarina, Catherine the Great in 1766.

His work on the motion of the Moon was of considerable interest to his sponsor since his lunar tables and his second lunar theory, published in 1772 under the title of *Theoria Motuum Lunae* in the Communications of Petropolis, helped the navigation of the Russian Navy. His lunar theory was also used, even before it appeared in its published form, by the Astronomer Royal Maskelyne in the British Nautical Almanac as the basis for the lunar ephemeris. This was first published in 1767 and was used by the British Navy for navigation. (These were probably the first, but certainly not the last uses of celestial mechanics by the military.)

His regularization in 1765 of the collision of straight-line motion is certainly one of the "firsts" in celestial mechanics. He will also be remembered from one of his results combining e, π and $i = \sqrt{-1}$ in the equation $e^{i\pi} = -1$. He is also credited with the method of variation of elements and with the introduction of moving coordinates in his lunar theory. His system was revolving but not rotating and, consequently, he did not find the Jacobian integral.

Several years before his death he became completely blind but his scientific activities did not stop; he dictated his papers, memoirs and scientific correspondence to his students.

Joseph Louis Lagrange (1736-1813) was born in Torino, Italy, where he was appointed professor of geometry at the artillery academy at the age of 19. In 1766 he went to Berlin, filling Euler's vacated position at the invitation of Frederick the Great, where he spent 20 years. The next invitation came from Louis XVI to Paris, where he became professor at the Ecole Polytechnique in 1797. His apartment in Paris was in the Louvre and he was buried in the Panthéon.

His announcement concerning the triangular libration points in the Sun-Jupiter system and his prediction of the possible existence of asteroids in these regions date from 1772. Observational astronomers could not verify the existence of these bodies for another 134 years. In this case, theory was certainly ahead of observations. His work on the solar system using the method of variation of parameters (1782) is one of the fundamental contributions in celestial mechanics.

Lagrange's celebrated *Méchanique Analytique* was published in 1788.

Marquis de **Pierre Simon Laplace** (1749-1827) was born in Beaumont-en-Auge and became professor at the Ecole Militaire in Paris at the age of 18. One of his major contributions was concerned with the stability of the solar system (1773, 1784), a problem still unsolved today. He introduced the concept of the potential function and what is known today as Laplace's equation (1785). His lunar theory, published in 1802, followed

Euler's. The five volumes of his *Mécanique Céleste* were published between 1799 and 1825.

He is accredited with the introduction of the expression "il est aisé à voir" (it is easy to see) which is still used to assign lengthy derivations to the readership.

Karl Friedrich Gauss (1777-1855) offered the basic principles of orbit determination, several of which (sometimes under different names) are still used today. His admiration of Newton's *Principia* began at the age of 17. At 30 he became director of the observatory and professor of mathematics at the University of Göttingen, which position he kept for the rest of his life. Needless to say, he did not like to travel, which definitely distinguishes him from other scientists.

Sir William Rowan Hamilton (1805-1865) was educated on Newton's publications when he was in his teens. He spent most of his life at the Dunsink Observatory near Dublin. *On a General Method in Dynamics* (published in 1834 and 1835) offered a new approach based on the principle of varying action. Hamilton's influence on celestial mechanics is related to the concept of generating functions, widely used in general perturbation problems. His *Elements of Quaternions* (1866) found use in one of the modern regularization techniques in celestial mechanics.

Henri Poincaré (1854-1912) was one of the most prolific writers in the field of mathematics and celestial mechanics, contributing more than 30 books and 500 memoirs. The three volumes of his *Méthodes Nouvelles de la Mécanique Céleste* appeared in 1892, 1893 and 1899 and were translated recently into English by NASA. This was followed by his *Leçons de Mécanique Céleste* in 1905-1910. Concentrating on the problem of three bodies, Poincaré established the concept of non-integrable dynamical systems. His theorem seriously affected the results of workers who intended to show the stability of the solar system by representing the orbital elements of the planets in Fourier series. Since these series in general are divergent according to Poincaré's theorem, the "solutions" do not show stability.

Poincaré lectured in Paris, proofreading simultaneously the text of his *Méthodes*. This technique certainly saved time but left a number of errors in the text. It is a challenge still today to read the *Méthodes*, which suggests many unsolved problems. The high quality of Poincaré's work is being continued in Paris at the Institute Poincaré.

François Félix Tisserand (1845-1896) was born in Nuits-Saint-Georges, Côte-d'Or and became one of the few astronomers who successfully combined observations with theoretical analysis. His four volume *Traité de Méchanique Céleste* which appeared in 1889-1896 can be read with considerable ease (as compared to the *Principia* or to the *Méthodes*) and with great benefit.

Forest Ray Moulton (1872-1952) is the author of the first book on celestial mechanics published in the United States (1902, Macmillan Company, New York). It served for many years as the basic textbook in the U.S. The following quotation might show prejudice but it will please the devotees of our field: "At the present time Celestial Mechanics is entitled to be regarded as the most perfect science and one of the most splendid achievements of the human mind."

George David Birkhoff's (1884-1944) analytical work contributed much to dynamics and to the problem of three bodies. A similarity between Birkhoff's and Poincaré's approaches to dynamics is unquestionably clear. His *Dynamical Systems* (1927) and *Collected Mathematical Papers* (1950) are responsible for many investigations appearing in the modern literature concerning periodic orbits, the restricted problem, etc.

Edmund Taylor Whittaker's (1873-1956) book *A Treatise on the Analytical Dynamics of Particles and Rigid Bodies* (1904) is one of the fundamental and widely used texts in advanced dynamics. His many references and examples concerning the restricted and the general problems of three bodies, his elegant treatments of generalized gravitational problems, his interpretation and explanation of Poincaré's work, his introduction to regularization, etc., strongly recommend his book and his other publications to the workers in celestial mechanics.

Dirk Brouwer (1902-1966) was the leader of Yale University's group in celestial mechanics before and during the time when the launching of the first artificial satellites called the scientific community's attention to the immediate importance of and need for celestial mechanics. Some of the principal teachers with Brouwer were Clemence, Eckert, Hagihara and Herget. The list of assistants and participating visitors is also of interest showing Brouwer's influence on the development of celestial mechanics. Some of the names are Aarseth, Arenstorf, Contopoulos, Danby, Davis, Deprit, Duncombe, Garfinkel, Hénon, Hori, Izsák, Kovalevsky, Kozai, Kyner, Lánczos, Lecar, Lundquist, Marchal, Message, Moser, Musen, Nahon, O'Keefe, Pines, Pollard, Rabe, Schubart, Simo, Siry, Szebehely, Vicente, Vinti and Waldvogel. And the list continues with some of the representatives of the next generation whose names are well known in today's celestial mechanics: Aksnes, Bettis, Bozis, Chapront, Cochran, de Vries, Eades, Ferraz-Mello, Fiala, Giacaglia, Howland, Jefferys, Junqueira, Jupp, Laubscher, Lieske, Marsden, Martin, Meffroy, Morando, Mulholland, Nacozy, O'Handley, Oesterwinter, Ollongren, Ovenden, Peters, Pierce, Richardson, Saari, Seidelmann, Standish, Van Flandern, Williams and Zare, just to mention a few.

Samuel Herrick's (1911-1974) school on the west coast of the U.S. concentrated more on space dynamics and on space applications of dynamics than the Yale group. His text, *Astrodynamics* (1971 and 1972), his universal variables (not far removed from the concept of regularized variables) and his methods of special perturbations made his school well known. Herrick is often credited with introducing the word astrodynamics. His work and his students made this expression popular among engineers and scientists. The word seems to appear first in R.A. Sampson's paper published in the memoirs of the Royal Astronomical Society, (Vol 63, p. 3, 1921) discussing the motion of Jupiter's satellites. His use of "astrodynamics" is a combination of astronomy and dynamics. The principal contributors and authors associated with Herrick were Baker, Escobal, Junkins, Makemson and Westrom.

G. N. Duboshin (1904-1986) was the author of several books and articles applying quantitative and qualitative dynamics to space research as well as to astronomy. He was the Chairman of the Department of Celestial Mechanics at Moscow University for 25 years with an honorary degree from Torun, birthplace of Copernicus. He was a cultured man and passed on to his students his high standards of scientific and general ethics. His group in Moscow and the Institute of Theoretical Astronomy in Leningrad made many basic advances in celestial mechanics. Abhyankar, Aksenov, Alekseev, Anasova, Arnold, Brumberg, Chebotarev, Chetayev, Demin, Dubyago, Egorov, Elyasberg, Faddeev, Fesenkov, Krylov, Merman, Moiseev, Orlov and Subbotin are some of the contributors. Kolmogorov's and Lyapunov's fundamental work in dynamics found many applications to celestial mechanics.

The reader interested in historical details will enjoy some of the books listed in the Appendix: Andrade (1954); Bate, Mueller and White (1971); Beer and Strand (1975); Koestler (1959) and Lerner (1973). In addition *Men of Mathematics* by E.T. Bell, Simon and Schuster Publ., New York (1937), *The Great Ideas Today* edited by R. M. Hutchins and M. J. Adler, Encyclopaedia Britannica, Inc. (1973), *From Galileo to Newton* by A. R. Hall, Dover Publ., New York (1981) and *The Space Station* by H. Mark, Duke University Press, Durham, North Carolina (1987) are recommended. Regarding nondeterministic dynamics and uncertainties in celestial mechanics, see J. Lighthill's "The Recently Recognized Failure of Predictability in Newtonian Dynamics," Proc. Roy. Soc., Vol. A407, pp. 35-50, 1986, and I. Prigogine's (1980) book listed in the Appendix.

For additional fascinating details of the early history see "Copernicus and Tycho" by O. Gingerich, Scientific American, Vol. 229, No. 6, pp. 86-101, 1973. For Newton's contributions to cosmology, see *The First Three Minutes* by S. Weinberg, Bantam Books Inc., N.Y. (1977).

CHRONOLOGICAL LIST OF MAJOR CONTRIBUTORS TO CELESTIAL MECHANICS			
Aristotle	384 BC - 322 BC		
Ptolemaeus, C.	100 - 178	Bruns, H.	1848 - 1919
Copernicus, N.	1473 - 1543	Poincaré, J.H.	1854 - 1912
Brahe, T.	1546 - 1601	Charlier, C.V.L.	1862 - 1934
Galilei, G.	1564 - 1642	Painlevé, P	1863 - 1933
Kepler, J.	1571 - 1630	Brown, E.W.	1866 - 1938
Descartes, R.	1596 - 1650	Burrau, C.	1867 - 1944
Newton, I.	1642 - 1727	Stromgren, E.	1870 - 1947
Leibnitz, G.W.	1646 - 1716	Cowell, P.H.	1870 - 1949
Halley, E.	1656 - 1742	De Sitter, W.	1872 - 1934
Euler, L.	1707 - 1783	Moulton, F.R.	1872 - 1952
Clairaut, A.C.	1713 - 1765	Levi-Civita, T.	1873 - 1941
D'Alembert, J.	1717 - 1783	Sundman, K.F.	1873 - 1949
Lambert, J.H.	1728 - 1777	Whittaker, E.T.	1873 - 1956
Lagrange, J.L.	1736 - 1813	Plummer, H.C.	1875 - 1946
Herschel, W.F.	1738 - 1822	Hohmann, W.	1880 - 1945
Bode, J.E.	1747 - 1826	Shook, G.A.	1882 - 1954
Laplace, P.S.	1749 - 1827	Birkhoff, G.D.	1884 - 1944
Legendre, A.M.	1752 - 1833	Smart, W.M.	1889 - 1975
Gauss, K.F.	1777 - 1855	Lemaitre, G.E.	1894 - 1966
Poisson, S.D.	1781 - 1840	Siegel, C.L.	1896 - 1981
Encke, J.F.	1791 - 1865	Hagihara, Y.	1897 - 1979
De Coriolis, G.G.	1792 - 1843	Moiseev, N.D.	1902 - 1955
Herschel, J.F.W.	1792 - 1871	Brouwer, D.	1902 - 1966
Hansen, P.A.	1795 - 1874	Eckert, W.J.	1902 - 1971
Jacobi, K.G.J.	1804 - 1851	Wintner, A.	1903 - 1958
Hamilton, W.R.	1805 - 1865	Kolmogorov, A.N.	1903 - 1987
Leverrier, U.J.J.	1811 - 1877	Duboshin, G.N.	1904 - 1986
Delauney, C.E.	1816 - 1872	Kuiper, G.P.	1905 - 1973
Adams, J.C.	1819 - 1892	Clemence, G.M.	1908 - 1974
Airy, D.	1835 - 1881	Herget, P.	1908 - 1981
Newcomb, S.	1835 - 1909	Stiefel, E.L.	1909 - 1987
Thiele, T.N.	1838 - 1910	Herrick, S.	1911 - 1974
Hill, G.W.	1838 - 1914	Chebotarev, G.A.	1913 - 1975
Tisserand, F.F.	1845 - 1896	Pollard, H.	1919 - 1985
Darwin, G.H.	1845 - 1912	Colombo, G.	1920 -1984

Chapter 3. The Problem of Two Bodies

In this chapter we will formulate the equations of motion of two bodies which interact gravitationally.

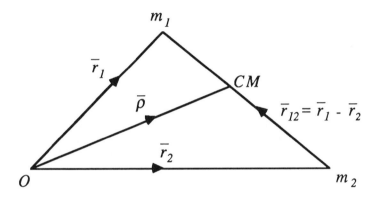

Figure 3.1. The Problem of Two Bodies.

Figure 3.1 shows two masses, m_1 and m_2, with their position vectors $\bar{r}_1$ and $\bar{r}_2$, which connect the origin of the inertial system of reference (O) with m_1 and m_2. The vector $\bar{\rho}$ gives the position of the center of mass of m_1 and m_2. Finally, the vector connecting m_1 and m_2 is $\bar{r}_{12} = \bar{r}_1 - \bar{r}_2$. The derivation of the equations of motion is rather straightforward since it is based on Newton's second law of motion and his law of gravity. According to this, the force acting on m_1 due to m_2 is equal to the time rate of change of the momentum of m_1, or $m_1\bar{r}_1$, which is equated to the gravitational force:

$$m_1\ddot{\bar{r}}_1 = -G\frac{m_1m_2}{|\bar{r}_{12}|^3}\bar{r}_{12}. \tag{3.1}$$

The force acting on m_2 is the same but opposite in direction. The equation of motion of m_2 becomes

$$m_2 \ddot{\bar{r}}_2 = G \frac{m_1 m_2}{|\bar{r}_{12}|} \bar{r}_{12} . \tag{3.2}$$

Note that we can cancel m_1 from the first equation and m_2 from the second equation. The two vector equations, (3.1) and (3.2), represent a system of six second order differential equations which is usually referred to as a twelfth order system of ordinary differential equations. (Both of these vector equations represent three scalar equations with second order derivatives.) Adding the above equations we obtain

$$m_1 \ddot{\bar{r}}_1 + m_2 \ddot{\bar{r}}_2 = 0. \tag{3.3}$$

Integrating this equation we have

$$m_1 \dot{\bar{r}} + m_2 \dot{\bar{r}}_2 = \bar{a} , \tag{3.4}$$

where $\bar{a}$ is a constant vector.

Once again integrating the above equation we obtain

$$m_1 \bar{r}_1 + m_2 \bar{r}_2 = \bar{a}t + \bar{b} , \tag{3.5}$$

where $\bar{b}$ is a constant vector. This last equation is often referred to as the integral of the motion of the center of mass. This is explained considering that the vector $\bar{\rho}$ shown on Figure 3.1 being the position vector of the center of mass, may be written as

$$\bar{\rho} = \frac{m_1 \bar{r}_1 + m_2 \bar{r}_2}{m_1 + m_2} . \tag{3.6}$$

Consequently, the motion of the center of mass is given by the equation

$$\bar{\rho} = \frac{\bar{a}t + \bar{b}}{m_1 + m_2} , \tag{3.7}$$

where $\bar{a}$ and $\bar{b}$ are constants of integrations and depend on the initial conditions.

From this we see that the center of mass is moving at a constant speed in a constant direction. Note that the center of mass might be stationary if the initial conditions are such that the vector $\bar{a}$ is zero.

Subtracting equation (3.2) from (3.1), and after cancelling the proper masses, we obtain

$$\ddot{\bar{r}}_{12} = -G\frac{(m_1 + m_2)}{|\bar{r}_{12}|^3}\bar{r}_{12}. \qquad (3.8)$$

This equation is the fundamental differential equation of the problem of two bodies written in terms of the vector connecting m_1 and m_2 which is known as the relative position vector. We may write this vector equation in terms of three scalar equations as

$$\ddot{x} = -G\frac{(m_1 + m_2)x}{(x^2 + y^2 + z^2)^{3/2}}, \qquad (3.9)$$

$$\ddot{y} = -G\frac{(m_1 + m_2)y}{(x^2 + y^2 + z^2)^{3/2}}, \qquad (3.10)$$

$$\ddot{z} = -G\frac{(m_1 + m_2)z}{(x^2 + y^2 + z^2)^{3/2}}, \qquad (3.11)$$

where x, y, z are the scalar components of $\bar{r}_{12}$.

The above three second order differential equations form a sixth order system. This system represents the relative motion between the two bodies. The sixth order system may be written as six first order differential equations, but this is not important for us at this point.

Two important results may be obtained from Eqn. (3.8), the conservation of energy and the conservation of angular momentum.

In the following derivations the relation

$$\bar{x} \cdot \bar{x} = |\bar{x}|^2 = (\bar{x})^2$$

is used, which follows from the definition of the scalar product of vectors. For instance

$$\frac{d}{dt}|\bar{x}|^2 = \frac{d}{dt}(\bar{x})^2 = \frac{d}{dt}(\bar{x} \cdot \bar{x}) = \bar{x} \cdot \dot{\bar{x}} + \dot{\bar{x}} \cdot \bar{x} = 2\bar{x} \cdot \dot{\bar{x}}.$$

Multiplying both sides of Eqn. (3.8) by the relative velocity vector using scalar or dot products one obtains

$$\dot{\bar{r}}_{12} \cdot \ddot{\bar{r}}_{12} = -G\frac{(m_1 + m_2)}{|\bar{r}_{12}|^3}\dot{\bar{r}}_{12} \cdot \bar{r}_{12}. \qquad (3.12)$$

From this equation one of the important equations of celestial mechanics, known as the equation of conservation of energy, might be obtained. The left-hand side can be written as one-half of the time derivative of the square of the velocity vector:

$$\frac{1}{2}\frac{d}{dt}(\dot{\bar{r}}_{12})^2 = \dot{\bar{r}}_{12} \cdot \ddot{\bar{r}}_{12}. \qquad (3.13)$$

The right-hand side may also be written as a time derivative, as shown below.

First we compute the time derivative of the quantity $1/|\bar{r}_{12}|$ using the chain rule:

$$\frac{d}{dt}\frac{1}{|\bar{r}_{12}|} = -\frac{1}{|\bar{r}_{12}|^2}\frac{d}{dt}|\bar{r}_{12}| \ . \tag{3.14}$$

The time derivative of $|\bar{r}_{12}|$ may be computed using the relation

$$|\bar{r}_{12}| = \left[\bar{r}_{12}^2\right]^{\frac{1}{2}} .$$

In this way we obtain

$$\frac{d}{dt}|\bar{r}_{12}| = \frac{d}{dt}\left(\bar{r}_{12}^2\right)^{\frac{1}{2}} = \frac{\bar{r}_{12}\cdot\dot{\bar{r}}_{12}}{\left[\bar{r}_{12}^2\right]^{\frac{1}{2}}} = \frac{\bar{r}_{12}\cdot\dot{\bar{r}}_{12}}{|\bar{r}_{12}|} , \tag{3.15}$$

and by substitution into Eqn. (3.14) we have

$$\frac{d}{dt}\frac{1}{|\bar{r}_{12}|} = -\frac{\bar{r}_{12}\cdot\dot{\bar{r}}_{12}}{|\bar{r}_{12}|^3} . \tag{3.16}$$

Consequently, our Eqn. (3.12) may be written as

$$\frac{1}{2}\frac{d(\dot{\bar{r}}_{12})^2}{dt} = \frac{d}{dt}\frac{G(m_1+m_2)}{|\bar{r}_{12}|} . \tag{3.17}$$

Integrating we obtain

$$(\dot{\bar{r}}_{12})^2 = \frac{2G(m_1+m_2)}{|\bar{r}_{12}|} + k , \tag{3.18}$$

where k is a constant of integration.

Writing $\bar{v} = \dot{\bar{r}}_{12}$ or $v^2 = |\dot{\bar{r}}|^2$ and $|\bar{r}_{12}| = r$, we obtain the equation of conservation of energy, also known as the energy integral, the energy equation, or the vis-viva integral:

$$v^2 = \frac{2GM}{r} + k , \tag{3.19}$$

where $M = m_1 + m_2$, v is the magnitude of the relative velocity, and r is the distance between the participating bodies.

Returning now to Eqn. (3.8) we multiply both sides by $\bar{r}_{12}$, but this time we use vectorial product, also known as cross product. Since the right-hand side of the equation is along the direction of $\bar{r}_{12}$ we obtain 0. Consequently, the result of this operation is that $\bar{r}_{12} \times \ddot{\bar{r}}_{12} = 0$. This equation means that the relative acceleration and the relative position vectors are along the same line.

Note also that the above cross product may be written as the time derivative of another cross product or

$$\ddot{\bar{r}}_{12} \times \bar{r}_{12} = \frac{d}{dt}(\dot{\bar{r}}_{12} \times \bar{r}_{12}).$$ (3.20)

The proof is obtained by computing the derivative of the product on the right hand side, realizing that $\dot{\bar{r}}_{12} \times \dot{\bar{r}}_{12} = 0$. Since the derivative on the right-hand side is zero, the quantity itself must be constant and consequently we arrive at the equation known as the conservation of the angular momentum:

$$\dot{\bar{r}}_{12} \times \bar{r}_{12} = \bar{c},$$ (3.21)

where $\bar{c}$ is the constant of angular momentum. Writing $\bar{r}$ for $\bar{r}_{12}$, we have

$$|\bar{c}| = |\dot{\bar{r}}_{12} \times \bar{r}_{12}| = |\bar{r} \times \dot{\bar{r}}| = |\bar{r}| \; |\dot{\bar{r}}| \sin\alpha,$$ (3.22)

where α is the angle between the directions of $\bar{r}$ and $\dot{\bar{r}}$, see Fig.3.2.

The components of the velocity vector $\dot{\bar{r}}$ are $\dot{r}$ and $r\dot{\phi}$, using polar coordinates. Note that $\dot{r}$ is the radial component of the velocity vector and it is not $|\dot{\bar{r}}|$. (On the other hand $|\bar{r}| = r$ is the length of the relative position vector.) (Eqn. 3.22) therefore, becomes

$$|\bar{c}| = r^2\dot{\phi}, \; or \; c = r^2\dot{\phi},$$ (3.23)

since $|\dot{\bar{r}}| \sin\alpha = r\dot{\phi}$.

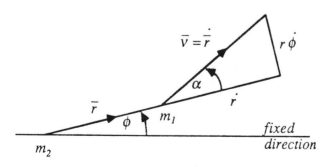

Figure 3.2. Polar Coordinates for the Problem of Two Bodies.

In Eqn. (3.23) and on Fig.3.2, ϕ is the angle between a fixed direction and the relative position vector $\bar{r}$, $r\dot{\phi}$ is the velocity component normal to $\bar{r}$ and c is the magnitude of the angular momentum vector.

The fact that the angular momentum vector is conserved means that the direction and the magnitude of the vector $\bar{c}$ in Eqn. (3.21) are constant. The direction of $\bar{c}$ is normal to the plane formed by $\bar{r}$ and $\dot{\bar{r}}$; therefore, the motion takes place in the fixed plane defined by the position and the velocity vectors.

The use of polar coordinates may simplify considerably the equations of celestial mechanics as seen when Eqn. (3.22) is compared to Eqn. (3.23).

The equation of the conservation of the energy (Eqn. 3.19) using polar coordinates becomes

$$v^2 = (r\dot{\phi})^2 + \dot{r}^2 = \frac{2\mu}{r} + k \, , \qquad (3.24)$$

where $\dot{r}$ and $r\dot{\phi}$ are the radial and normal velocity components as mentioned before and $\mu = G(m_1 + m_2)$. It should be noted that the symbol μ is also used for another well-known constant in celestial mechanics which is related to the mass ratio of the primaries in the restricted problem of three bodies, to be discussed later.

These conservation principles are not surprising. There is no dissipation in the system, consequently the system is conservative, and we expect the energy to be conserved. Furthermore, there is no outside moment acting on the system since the only force under consideration is the gravity force which is a so-called central force acting along the relative position vector $\bar{r}$. Consequently, the angular momentum is also conserved.

Before demonstrating the importance and practical usefulness of these conservation laws, the equation of motion of the problem of two bodies is represented using polar coordinates.

The variables are r and ϕ, instead of x, y, and z. Note that the motion always takes place in a plane (because of the conservation of the angular momentum); therefore, the two-dimensional version of Eqn. (3.8) is satisfactory. The two-dimensional equations may be obtained from Eqn. (3.8) or from Eqns. (3.9) - (3.11) by writing $z = 0$. In this way we have

$$\ddot{x} = -\frac{\mu x}{(x^2 + y^2)^{3/2}} \, , \qquad (3.25)$$

$$\ddot{y} = -\frac{\mu y}{(x^2 + y^2)^{3/2}} \, . \qquad (3.26)$$

The corresponding equations using polar coordinates may be obtained in several ways. One of the conventional methods is to use the equations of transformation from polar to Cartesian rectangular coordinates, ($x = r \cos\phi$, $y = r \sin\phi$), taking derivatives and substituting in Eqns. (3.25) and (3.26). This exercise is left to those readers who enjoy analytical manipulations. A more direct approach is to consider the fact that the radial acceleration consists of two terms, the second time derivative of the radial coordinate ($\ddot{r}$) and the centripetal acceleration ($-r\dot{\phi}^2$). Equating these accelerations with the force in the radial direction which is the gravitational force ($-\mu/r^2$), we have

$$\ddot{r} - r\dot{\phi}^2 = -\frac{\mu}{r^2}. \tag{3.27}$$

The differential equation of motion for the angular coordinate ϕ may be obtained by taking the time derivative of the equation representing the conservation of the angular momentum, $r^2\dot{\phi} = c$. In this way we obtain

$$2\dot{r}\dot{\phi} + r\ddot{\phi} = 0, \tag{3.28}$$

which equation shows that the only force acting is in the radial direction.

So the two second order differential equations of motion are represented by Eqns. (3.27) and (3.28), one representing the radial and the other the normal component. From the angular momentum conservation equation we may express $\dot{\phi}$ as c/r^2 and substitute this in the radial equation of motion. In this way the angular coordinate is eliminated, and we obtain

$$\ddot{r} - c^2/r^3 = -\mu/r^2. \tag{3.29}$$

The new equation (3.29) of the problem of two bodies is a second order nonlinear differential equation representing the variation of the distance with time. If this differential equation is solved and the solution $r(t)$ is substituted into Eqn. (3.23) connecting ϕ and r, the variation of the angular coordinate ϕ with time may be obtained:

$$\phi = \int \frac{c}{r^2} dt. \tag{3.30}$$

In this way the complete solution of the problem of two bodies will be given by $r(t)$ and $\phi(t)$. As we will see this process can be followed only in principle since one of the challenges of celestial mechanics is that the solution of the simplest problem known as the problem of two bodies cannot be expressed by a closed form, explicit function of the time.

Another remark might be appropriate at this time. In order to find the absolute motions of m_1 and m_2 in addition to their relative motion, we might attempt to solve Eqn. (3.8) and obtain $\bar{r}_{12}$ as a function of time. The previous paragraph denies the existence of a closed form analytical expression for this problem; nevertheless, as we will see later, such solutions might be obtained by using different variables. The transition from $\bar{r}_{12}$ to $\bar{r}_1$ and $\bar{r}_2$ is still required in order to have information concerning the motions of m_1 and m_2. This might be obtained using Eqn. (3.6) and Fig. 3.1 as follows. The vectors $\bar{\rho}$ and $\bar{r}_{12}$ are linear functions of the position vectors $\bar{r}_{12}$ and $\bar{r}_2$ as we have seen before:

$$\bar{r}_{12} = \bar{r}_1 - \bar{r}_2$$

and

$$\bar{\rho} = \frac{m_1\bar{r}_1 + m_2\bar{r}_2}{m_1 + m_2}.$$

From these equations $\bar{r}_1$ and $\bar{r}_2$ are obtained as

$$\bar{r}_1 = \bar{\rho} + \frac{m_2}{M}\bar{r}_{12},$$

$$\bar{r}_2 = \bar{\rho} - \frac{m_1}{M}\bar{r}_{12}.$$

If m_2 is much larger than m_1, we have $M \approx m_2$, $m_1 \approx 0$ and

$$\bar{r}_1 \approx \bar{\rho} + \bar{r}_{12},$$

$$\bar{r}_2 \approx \bar{\rho}.$$

In this case, therefore, the center of mass coincides with m_2.

This chapter is concluded with the remark that the problem of two bodies may be represented by Eqns. (3.25) and (3.26) forming a fourth order system or by Eqns. (3.23) and (3.29) forming a third order system, or by Eqns. (3.23) and (3.24) forming a second order system, or finally by Eqn. (3.29), which is a second order differential equation.

In the following it will be shown how the variables of these equations may be changed so that a closed form exact analytical solution is obtained. Using the "proper" independent and dependent variables our nonlinear system of differential equations may be transformed into a linear differential equation describing a harmonic oscillator. In this way the challenge of celestial mechanics will be converted into a problem of transformations instead of a problem of solving nonlinear differential equations. Nevertheless, any of the above system might be subjected to numerical integrations, giving results useful for some specific cases. The general behavior of our dynamical system - the gravitational problem of two bodies - however, can not be understood from such special examples.

Introductory chapters from Danby's (1962) and Roy's (1978) books are recommended.

EXAMPLES

1. Derivation of the equations of motion using polar coordinates might be performed by transformations of Eqns. (3.25) and (3.26) from x, y, to r, ϕ. The equations of the transformation are

$$x = r\cos\phi \text{ and } y = r\sin\phi,$$

from which by taking time derivatives we obtain

$$\ddot{x} = \ddot{r}\cos\phi - r\ddot{\phi}\sin\phi - 2\dot{r}\dot{\phi}\sin\phi - r\dot{\phi}^2\cos\phi$$

and

$$\ddot{y} = \ddot{r}\sin\phi + r\ddot{\phi}\cos\phi + 2\dot{r}\dot{\phi}\cos\phi - r\dot{\phi}^2\sin\phi.$$

Substituting in Eqns. (3.25) and (3.26) we obtain

$$(\ddot{r} - r\dot{\phi}^2)\cos\phi - (r\ddot{\phi} + 2\dot{r}\dot{\phi})\sin\phi = -\frac{\mu\cos\phi}{r^2}$$

$$(\ddot{r} - r\dot{\phi}^2)\sin\phi + (r\ddot{\phi} + 2\dot{r}\dot{\phi})\cos\phi = -\frac{\mu\sin\phi}{r^2}.$$

Multiplying the first equation by $\cos\phi$, the second by $\sin\phi$ and adding, we have

$$\ddot{r} - r\dot{\phi}^2 = -\frac{\mu}{r^2}.$$

Multiplying the first equation by $\sin\phi$, the second by $\cos\phi$ and subtracting, we have

$$r\ddot{\phi} + 2\dot{r}\dot{\phi} = 0.$$

These last two equations correspond to Eqns. (3.27) and (3.28).

2. For circular motion r is constant, say $r = r_o$. Eqn. (3.27) gives

$$r_o\dot{\phi}^2 = \frac{\mu}{r_o^2}$$

and from Eqn. (3.28) we have $\ddot{\phi}r_o = 0$, that is, $\dot{\phi}$ is a constant, $\dot{\phi} = \dot{\phi}_o$. Substituting this in the above equation, gives

$$r_o\dot{\phi}_o^2 = \frac{\mu}{r_o^2}$$

of $r_o^3 \dot{\phi}_o^2 = \mu$ which is equivalent to Eqn. (1.16) since $\dot{\phi}_o$ is the angular velocity or mean motion, previously denoted by n.

3. The solution of the problem of two bodies can be represented by Taylor-series. This solution is known in the literature as the f and g series.

We write the solution as

$$\bar{r} = \bar{r}_o f + \dot{\bar{r}}_o g,$$

where the functions f and g depend on the time and on the initial conditions.

The Taylor-series solution is

$$\bar{r} = \bar{r}_o + \dot{\bar{r}}_o (t - t_o) + \frac{\ddot{\bar{r}}_o}{2}(t - t_o)^2 + \frac{\dddot{\bar{r}}_o}{3!}(t - t_o)^3 + \cdots .$$

The initial conditions are represented by the position and velocity vectors at $t = t_o$, which are denoted by $\bar{r}_o$ and $\dot{\bar{r}}_o$. All other coefficients of the series can be expressed by these vectors as shown by the following derivation. From the equation of motion we have

$$\ddot{\bar{r}} = -\mu \frac{\bar{r}}{|\bar{r}|^3},$$

corresponding to Eqn. (3.8) with $\bar{r} = \bar{r}_{12}$ and $\mu = G(m_1 + m_2)$.

At $t = t_o$, the above equation becomes

$$\ddot{\bar{r}}_o = -\mu \frac{\bar{r}_o}{|\bar{r}_o|^3}.$$

In this way the third term of the series is obtained. The fourth term requires the computation of the third derivative of $\bar{r}$ from the equation of motion:

$$\dddot{\bar{r}} = -\mu \frac{\dot{\bar{r}}|\bar{r}|^3 - \bar{r}\dfrac{d|\bar{r}|^3}{dt}}{|\bar{r}|^6}.$$

Here the derivative of $|\bar{r}|^3$ requires some attention but its evaluation is not more difficult than the previously discussed derivative of $|\bar{r}_{12}|^{-1}$. We have

$$\frac{d|\bar{r}|^3}{dt} = \frac{d(\bar{r}^2)^{3/2}}{dt} = \frac{3}{2}\sqrt{\bar{r}^2}\frac{d\bar{r}^2}{dt} = \frac{3}{2}|\bar{r}|\, 2\bar{r} \cdot \dot{\bar{r}}.$$

The third derivative becomes

$$\dddot{r} = -\mu \frac{\dot{r}|\bar{r}|^3 - \bar{r}\, 3(\bar{r} \cdot \dot{r})|\bar{r}|}{|\bar{r}|^6}$$

or

$$\dddot{r} = -\mu \frac{\dot{r}}{|\bar{r}|^3} + 3\mu(\bar{r} \cdot \dot{r})\frac{\bar{r}}{|\bar{r}|^5} .$$

At $t = t_o$ we have

$$\dddot{r}_o = -\mu \frac{\dot{r}_o}{|\bar{r}_o|^3} + 3\mu(\bar{r}_o \cdot \dot{r}_o)\frac{\bar{r}_o}{|\bar{r}_o|^5} .$$

The Taylor-series expansion becomes

$$\bar{r} = \bar{r}_o + \dot{r}_o(t - t_o) - \frac{\mu}{2} \frac{\bar{r}_o}{|\bar{r}_o|^3}(t - t_o)^2 + \left[\frac{\mu}{2}(\bar{r}_o \cdot \dot{r}_o)\frac{\bar{r}_o}{|\bar{r}_o|^5} \right.$$

$$\left. - \frac{\mu}{6} \frac{\dot{r}_o}{|\bar{r}_o|^3} \right](t - t_o)^3 + \cdots .$$

The members of the f series have the factor $\bar{r}_o$ and those of the g series are multiplied by $\dot{r}_o$:

$$f = 1 - \frac{\mu}{2} \frac{(t - t_o)^2}{r_o^3} + \frac{\mu}{2} \frac{(\bar{r}_o \cdot \dot{r}_o)}{r_o^5}(t - t_o)^3 \pm \cdots$$

$$g = t - t_o - \frac{\mu}{6} \frac{(t - t_o)^3}{r_o^3} \pm \cdots ,$$

where $r_o = |\bar{r}_o|$ is the length of the initial position vector.

Note that the second factor of the third term of the f series can be written as

$$\bar{r}_o \cdot \dot{r}_o = |\bar{r}_o||\dot{r}_o| \cos\alpha = r_o \dot{r}_o ,$$

where $\dot{r}_o = |\dot{r}_o| \cos\alpha$ is the initial value of the radial velocity component (see Fig. 3.2). This notation can be misleading since $|\bar{r}_o| = r_o$ is the length of the initial radial position vector, but $|\dot{r}_o| \neq \dot{r}_o$ as mentioned before.

With this notation the f series becomes

$$f = 1 - \frac{\mu}{2} \frac{(t - t_o)^2}{r_o^3} + \frac{\mu}{2} \frac{\dot{r}_o}{r_o^4}(t - t_o)^3 \pm \cdots .$$

For short times these series are the basic tools of orbit determination. The convergence properties are discussed in Chapter 8 of Taff's book, see Appendix. The first 35 terms of the series are given in A. Deprit's "Fundamentals of Astrodynamics." Mathematical Note No. 556, Boeing Scientific Research Laboratories, 1968.

PROBLEMS

1. Find the approximate distances between the Sun and the centers of mass of the Sun-Jupiter, Sun-Saturn and Sun-Earth systems. Find the approximate circular velocity of the center of the Sun in these systems.

2. Find the approximate distance between the Earth and the center of mass of the Earth-Moon system and find the approximate circular velocity of the center of the Earth in this system.

Chapter 4. Elliptic Orbits

In this chapter the geometry and dynamics of elliptic two-body orbits are discussed. The reader will notice that some of the results also apply to hyperbolic and parabolic orbits, which will be treated in more detail in Chapter 7.

The polar angle ϕ is related to the true anomaly f in celestial mechanics by the equation $\phi = f + \omega$ and it will now be introduced as the new independent variable. Here $\omega = \phi_o$ is the argument of pericenter, see Figure 4.1. Using the chain rule and Eqn. (3.23) we have

$$\frac{dr}{dt} = \frac{dr}{d\phi}\frac{d\phi}{dt} = \frac{c}{r^2}\frac{dr}{d\phi} = \frac{c}{r^2}r'.$$

The second derivative becomes

$$\frac{d^2r}{dt^2} = \frac{c^2}{r^4}(r'' - 2\frac{r'^2}{r}).$$

In these equations the notations

$$r' = \frac{dr}{d\phi} \quad \text{and} \quad r'' = \frac{d^2r}{d\phi^2}$$

are used.

Substituting in Eqn. (3.29) we obtain the new equation of motion using ϕ as the new independent variable:

$$r'' - 2\frac{r'^2}{r} - r = -\frac{\mu}{c^2}r^2. \tag{4.1}$$

This equation becomes the equation of a harmonic oscillator if the substitution $r = 1/u$ is made. We have

$$r' = -\frac{u'}{u^2} \quad \text{and} \quad r'' = \frac{2u'^2 - uu''}{u^3}.$$

Using the dependent variable u, instead of r, Eqn. (4.1) becomes

$$u'' + u = \mu/c^2. \tag{4.2}$$

The solution of this linear differential equation is well known and it may be written as

$$u = \frac{\mu}{c^2} + A \cos(\phi - \phi_o), \qquad (4.3)$$

where A and ϕ_o are constants of integration depending on the initial conditions. The radial distance between the bodies is the reciprocal of u, or

$$r = \frac{c^2/\mu}{1 + (Ac^2/\mu) \cos(\phi - \phi_o)} , \qquad (4.4)$$

which is the equation of a conic section, using polar coordinates. In the conventional notation Eqn. (4.4) becomes

$$r = \frac{p}{1 + e \cos(\phi - \phi_o)} , \qquad (4.5)$$

where p is the semi-latus rectum, e is the eccentricity and ϕ_o is zero if the true anomaly ϕ is measured from the pericenter. The eccentricity is less than 1 for elliptic orbits, larger than 1 for hyperbolas and $e = 1$ for parabolas. Comparing Eqns. (4.4) and (4.5) we have

$$p = c^2/\mu \text{ and } e = Ac^2/\mu . \qquad (4.6)$$

Here the constant of integration A is still to be related to the geometry and to the dynamics of the problem.

We found previously a form of the conservation of energy

$$v^2 = \dot{r}^2 + (r\dot{\phi})^2 = \frac{2\mu}{r} + k ,$$

where $\dot{\phi} = c/r^2$.

Using the new independent variable, this becomes

$$\frac{c^2}{r^4} r'^2 + \frac{c^2}{r^2} = \frac{2\mu}{r} + k. \qquad (4.7)$$

From Eqn. (4.5) we have

$$r' = \frac{ep \sin f}{(1+e \cos f)^2} , \qquad (4.8)$$

where $f = \phi - \phi_o$. The energy equation (4.7) after substituting r from Eqn. (4.5) and r' from Eqn. (4.8) results in

$$e = \sqrt{1 + kc^2/\mu^2} . \qquad (4.9)$$

If the energy constant k is positive, $e > 1$ and the orbit is a hyperbola. If $k < 0$, $e < 1$ and we have elliptic orbits. Finally, when $k = 0$, $e = 1$

represents parabolic orbits.

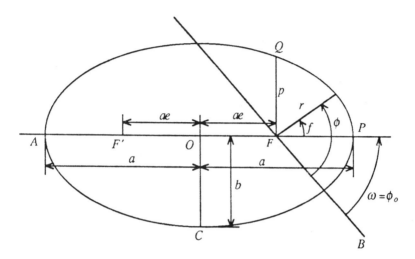

Figure 4.1. Elliptic Orbit.

In the following we shall discuss elliptic orbits in considerable detail and treat hyperbolic and parabolic orbits in Chapter 7, where straight-line orbits for which $e = 0$ and $c = 0$, will receive special attention. From Fig. 4.1 and Eqn. (4.5) we may see that

when $f = 0, r = p/(1 + e)$ and when $f = 180°, r = p/(1 - e)$.

Therefore, the distance between points P and A is

$$\frac{p}{1+e} + \frac{p}{1-e} = \frac{2p}{1-e^2} = 2a,$$

which relates the semi-latus rectum p and the semi-major axis a as

$$p = a(1 - e^2). \tag{4.10}$$

The semi-major axis is sometimes called the major semi axis in the British literature. It is often referred to as the mean distance between the focus and the elliptic orbit. (The careful interpretation of the mean or average distance is given in Example 4 at the end of Chapter 5.)

The point P is called the perigee for Earth-orbiting bodies (the Moon or artificial satellites); it is called the perihelion for Sun-orbiting bodies (planets, comets, asteroids, space probes). It is generally called pericenter, and in stellar dynamics it is known as the periastron. Point A has the corresponding terminology as apogee, aphelion, apocenter and apoastron. The pericenter is the closest and the apocenter is the farthest point from the focus F on the elliptic orbit.

The direction FB is considered fixed and the previously mentioned polar angle $\phi = f + \omega$ is measured from this direction. The line OP is the semi-major axis and its direction is given by $\phi_o = \omega$, known as the argument of the pericenter. Note that the direction of FB is not taken arbitrarily but it is along the line of nodes which will be discussed at the beginning of Chapter 9. The semi-minor axis is $\overline{OC} = b = a\sqrt{1 - e^2}$.

The pericenter distance (FP) is

$$r^P = \frac{p}{1 + e} = a(1 - e)$$

and the apocenter distance is

$$r_A = \frac{p}{1 - e} = a(1 + e).$$

As the next step we shall relate the constant of energy k to the geometry of orbit and we shall show that

$$k = -\frac{\mu}{a} \tag{4.11}$$

Note that the total energy is negative for elliptic orbits. We have from Eqn. (4.9)

$$k = (e^2 - 1)\frac{\mu^2}{c^2},$$

from Eqn. (4.10)

$$e^2 - 1 = -p/a,$$

and from Eqn. (4.6)

$$p = c^2/\mu.$$

The combination of these results gives $k = -\mu/a$. Note that there is another, more direct way to obtain the above result by making use of the conservation of energy and of angular momentum. The energy equation (3.24) gives the velocities at pericenter (v_P) and at apocenter (v_A):

$$v_P^2 = \frac{2\mu}{a(1-e)} + k,$$

$$v_A^2 = \frac{2\mu}{a(1+e)} + k.$$

From the conservation of angular momentum we have

$$v_A r_A = v_P r_P,$$

since $\bar{v}_A$ is perpendicular to $\bar{r}_A$ and $\bar{v}_P$ is perpendicular to $\bar{r}_P$.

Squaring both sides and substituting the above values of v_A^2, v_P^2, r_A and r_P we have

$$\left[\frac{2\mu}{a(1+e)} + k\right] a^2(1+e)^2 = \left[\frac{2\mu}{a(1-e)} + k\right] a^2(1-e)^2.$$

From this we obtain $k = -\mu/a$.

The energy equation for elliptic orbits is

$$v^2 = \frac{2\mu}{r} + k = \frac{2\mu}{r} - \frac{\mu}{a} = \mu(\frac{2}{r} - \frac{1}{a}). \tag{4.12}$$

Now we may compute new forms of the pericenter and apocenter velocities :

$$v_P^2 = \mu(\frac{2}{r_P} - \frac{1}{a}),$$

$$v_A^2 = \mu(\frac{2}{r_A} - \frac{1}{a}),$$

or

$$v_P^2 = \frac{\mu}{a}\frac{1+e}{1-e},$$

$$\tag{4.13}$$

$$v_A^2 = \frac{\mu}{a}\frac{1-e}{1+e}.$$

From these we obtain

$$v_P = \frac{1+e}{1-e}v_A. \tag{4.14}$$

For small values of the eccentricity we might use an approximation

$$v_P = \frac{1+e}{1-e}v_A \approx (1+e)^2 v_A \approx (1+2e)v_A.$$

For instance for $e = 0.05$ the pericenter velocity is 10% higher than the velocity at the apocenter. For large eccentricity, this approximation can not be used. For instance when $e = 0.9$, the velocity at the pericenter is 19 times the velocity at the apocenter. The two velocities become equal when $e = 0$, corresponding to circular orbits. For this case the previously derived Eqn. (1.5) gives

$$v_c = \left[\frac{\mu}{r} \right]^{\frac{1}{2}}.$$

The same result may be obtained from the energy conservation equation (4.12) by writing $r = a$ or from Eqns. (4.13) with $e = 0$.

Introductory chapters from Danby's (1962), McCuskey's (1963) and Roy's (1978) books are recommended.

EXAMPLES

1. Derive the radial $(\dot{r})$ and normal to radial $(r\dot{\phi})$ velocity components on elliptic orbits as functions of μ, a, e and ϕ, with $\phi_o = 0$.

    Using

    $$r = \frac{p}{1 + e\cos\phi} \quad \text{and} \quad r^2\dot{\phi} = \sqrt{\mu p} ,$$

    where p is the semi-latus rectum $p = a(1 = e^2)$, the radial velocity becomes

    $$\dot{r} = \frac{dr}{d\phi}\dot{\phi} = e\left[\frac{\mu}{a(1-e^2)} \right]^{\frac{1}{2}} \sin\phi .$$

    Note that for a circular orbit $e = 0$ and the radial velocity becomes zero, as expected.

    The normal velocity component is

    $$r\dot{\phi} = \frac{\sqrt{\mu p}}{r} = \left[\frac{\mu}{a(1-e^2)} \right]^{\frac{1}{2}} (1 + e\cos\phi) .$$

    For a circular orbit the normal velocity component becomes the circular velocity,

$$r\dot\phi = \left[\frac{\mu}{a}\right]^{1/2} = v_c .$$

The total velocity on elliptic orbits is

$$v = \left[\dot r^2 + r^2\dot\phi^2\right]^{1/2} = \left[\frac{\mu}{a(1-e^2)}\right]^{1/2}\left[1 + e^2 + 2e\cos\phi\right]^{1/2} ,$$

which again becomes the circular velocity for $e = 0$. The same result can be obtained from the integral of energy,

$$v^2 = \mu(\frac{2}{r} - \frac{1}{a}) ,$$

when the relation between r and ϕ is substituted.

2. The period, apogee height and perigee height of satellite Explorer 7 were 1.684 hours, 664 miles and 346 miles. Find the semi-major axis, semi-minor axis, semi-latus rectum in km, the perigee and apogee velocities in km/sec and the eccentricity of the orbit.

   Twice the value of the semi-major axis is the sum of the perigee and apogee distances, measured from the center of the Earth. The corresponding formula is

$$2a = 2R_E + h_P + h_A ,$$

where $h_P = 556.834\ km$, $h_A = 1068.606\ km$ and $R_E = 6378.14\ km$. The semi-major axis becomes $a = 7190.86\ km$.

   The eccentricity can be computed from the formula

$$e = \frac{h_A - h_P}{2a} .$$

This equation can be obtained using the definitions of the perigee and apogee distances :

$$r_P = h_P + R_E = a(1 - e) ,$$
$$r_A = h_A + R_E = a(1 + e) ,$$

and computing

$$h_A - h_P = a(1 + e) - a(1 - e) = 2ae .$$

The eccentricity becomes $e = 0.03558$.

   The semi-minor axis and the semi-latus rectum are given by

$$b = a\sqrt{1 - e^2}\ \text{and}\ p = a(1 - e^2).$$

The values are $b = 7186.306\ km$ and $p = 7181.757\ km$.

The perigee and apogee velocities are given by Eqns. (4.13) :

$$v_P = \left[\frac{\mu}{a}\right]^{1/2} \left[\frac{1+e}{1-e}\right]^{1/2} = 7.715 \ km/sec \ ,$$

$$v_A = \left[\frac{\mu}{a}\right]^{1/2} \left[\frac{1-e}{1+e}\right]^{1/2} = 7.185 \ km/sec \ .$$

3. Show that for elliptic orbits the semi-major axis (a) is larger than the semi-major axis (b), which in turn is larger than the semi-latus rectum (p), which finally is larger than the distance between the focus and the pericenter (r_P).

    The series of inequalities stated above can be written as

    $$a \geq b \geq p \geq r_P \ .$$

Here $b = a\sqrt{1-e^2}, p = a(1-e^2), r_P = a(1-e)$, and $0 \leq e \leq 1$.

The inequalities become

$$a \geq a\sqrt{1-e^2} \geq a(1-e^2) \geq a(1-e) \ .$$

The first inequality, after a is cancelled, becomes

$$1 \geq 1 - e^2 \ or \ 0 \geq -e^2 \ .$$

This last inequality becomes an equality for $e = 0$. The second inequality becomes

$$1 \geq \sqrt{(1-e^2)} \ or \ 1 \geq 1 - e^2 \ or \ 0 \geq -e^2 \ ,$$

after cancellation by $a\sqrt{1-e^2}$ and squaring. This again becomes an equality when $e = 0$. The third inequality becomes

$$1 + e \geq 1 \ or \ e \geq 0.$$

We conclude that the inequalities described are correct for elliptic orbits and become equalities for circular orbits. For straight-line or rectilinear orbits ($e = 1$) we have $b = p = r_P = 0$. (See Chapter 7, following Eqn. (7.8).

PROBLEMS

1. The perigee and apogee altitude of an artificial Earth satellite are 200 km and 500 km. Find the values of a, b, p, e, v_p and v_A.

2. Show that the maximum value of $\dot{r}$ for an elliptic orbit occurs at the intersection of the latus rectum with the orbit and find this maximum value.

Chapter 5. Kepler's Laws, Kepler's Equation and Regularization

Kepler's Laws.

The three laws Kepler offered will be reviewed and his equation connecting time with the eccentric anomaly will be derived in this chapter. Kepler's and Newton's work represents the foundations of celestial mechanics and the thorough understanding of these laws is essential in our field.

Kepler's laws deal with the kinematics of planetary orbits and offer descriptions of the motions rather than explanations. These laws might be summarized as follows:

1. The orbits of the planets are ellipses with the Sun at the focus (heliocentric solar system).

2. Law of areas: equal areas in equal times are covered by the radius vector connecting the Sun and a planet.

3. The squares of the periods of the planets are proportional to the cubes of their semi-major axes.

These laws follow from our previous discussions of the problem of two bodies. The first law (sometimes considered the second) modified Copernicus' circular planetary orbits, and for us it is an expression of the fact that for negative energy the solution of the two-body problem is an elliptic orbit (Eqn. 4.11). The second law, also known as the law of areas, expresses the conservation of the angular momentum since the area covered by the radius vector in time dt is

$$dA = (r^2 d\phi)/2$$

or

$$2\frac{dA}{dt} = r^2\dot{\phi} = c \, .$$

These relations were given by Eqns. (3.22) - (3.24) and (3.27). Figure

5.1 shows the relation between the area and the angular momentum.

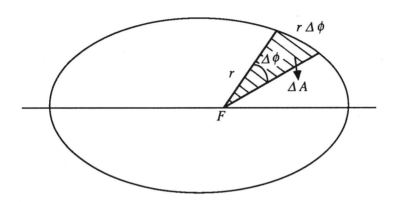

Figure 5.1. The Law of Areas.

The third law follows from the second by applying it to a complete period, $T = 2\pi/n$. The corresponding area of the ellipse is $A = ab\pi$, and therefore

$$\frac{dA}{dt} = \frac{ab\pi}{2\pi}\, n = \frac{abn}{2} \ ,$$

or

$$c = na^2\sqrt{1 - e^2}\ .$$

On the other hand, we have seen that the angular momentum may be expressed as $c = v_P a\,(1 - e)$ or as

$$c = \sqrt{\mu a}\ \sqrt{1 - e^2}\ .$$

Equating the above two expressions for c, we have

$$n^2 a^3 = \mu\ ,$$

which result was already given as Eqn. (1.16) for circular motion.

Consider now m_1 representing the mass of the Sun and m_2 the mass of a planet. Then we have

$$\frac{a^3}{T^2} = \frac{G(m_S + m_P)}{4\pi^2} \; .$$

Applying this equation to two planets with mass m_{P1} and m_{P2} we have

$$\frac{a_1^3}{T_1^2} = \frac{G(m_S + m_{P1})}{4\pi^2}$$

and

$$\frac{a_2^3}{T_2^2} = \frac{G(m_S + m_{P2})}{4\pi^2} \; .$$

The ratio becomes

$$\frac{T_1^2}{T_2^2} = \frac{a_1^3}{a_2^3} \frac{1 + m_{P2}/m_S}{1 + m_{P1}/m_S} \; . \tag{5.1}$$

Using this notation, Kepler's third law states that

$$\frac{T_1^2}{T_2^2} = \frac{a_1^3}{a_2^3} \; , \tag{5.2}$$

which is correct if the planetary masses are much smaller than the mass of the Sun. For the largest planet (Jupiter) the correction becomes 0.1%.

Kepler's Equation.

Kepler's equation relates the geometry of elliptic orbits to kinematics by relating the variables to time. The derivation starts with Eqns. (3.23) and (4.4). From the first we obtain

$$\int_{\phi_o}^{\phi} r^2 d\phi = c(t - t_o) \; ,$$

which might be modified by substituting the polar equation of an ellipse given by Eqn. (4.4). The limits of the integration connect the time of perigee passage t_o with ϕ_o or with $f = 0$, where $f = \phi - \phi_o$. After the parameters are evaluated using $c = \sqrt{p\mu}$ and $n^2 a^3 = \mu$, we have

$$n(t - t_o) = (1 - e^2)^{3/2} \int_0^f \frac{df}{(1 + e \cos f)^2} \; .$$

The right-hand side may be transformed to a rational algebraic function by the substitution $u = \tan(f/2)$. The result of the integration is

$$n(t - t_o) = -\frac{e\sqrt{1 - e^2}\sin f}{1 + e\,\cos f} + 2\arctan\left[\left[\frac{1 - e}{1 + e}\right]^{\frac{1}{2}}\tan(f/2)\right].\quad (5.3)$$

The result of the derivation, Eqn. (5.3) gives the mean anomaly $l = n(t - t_o)$ as a function of the true anomaly f. Kepler's equation relating the mean anomaly to the eccentric anomaly (E) is obtained from Eqn. (5.3) when the relation

$$r = a(1 - e\,\cos E) = p/(1 + e\,\cos f)\qquad (5.4)$$

is introduced. Eqn. (5.4) connects the true and eccentric anomalies and this allows us to express the right side of Eqn. (5.3) as a function of E instead of as a function of f. Before giving the geometrical meaning of the variable E, we complete the transformation of Eqn. (5.3). Note that this is straightforward using simple trigonometric relations such as

$$\cos f = \frac{\cos E - e}{1 - e\,\cos E} \text{ and } \tan\frac{f}{2} = \left[\frac{1 + e}{1 - e}\right]^{1/2}\tan\frac{E}{2},$$

which follow from Eqn. (5.4). The transformed Eqn. (5.3) becomes

$$n(t - t_o) = E - e\,\sin E,\qquad (5.5)$$

which is known as Kepler's equation.

The Anomalies.

Figure 5.2 shows a seldom used method of constructing an ellipse. Consider two concentric circles with radii a and b. Projecting the intersections of line $\overline{OC}$ with these circles (points B and C) parallel to a and parallel to b, a point of the ellipse (Q) is obtained. The line $\overline{FQ}$ represents r which might be obtained as

$$r = \sqrt{\overline{QD}^2 + \overline{FD}^2},$$

or

$$r = \sqrt{(b\sin E)^2 + (a\,\cos E - ea)^2},$$

from which Eqn. (5.4) is obtained.

The geometrical meaning of the eccentric anomaly is shown on Figure 5.2 as the central angle E.

Note that Kepler's equation may also be obtained by introducing

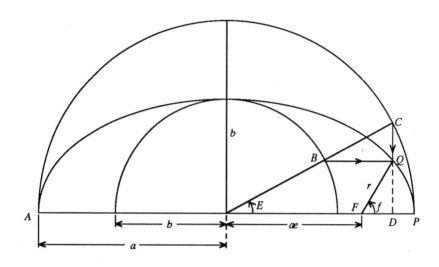

Figure 5.2. True and Eccentric Anomalies.

the eccentric anomaly before integrating the equation of the angular momentum. In this way we have from $r^2\dot\phi = c$,

$$\frac{n}{c}\int_{\phi_o}^{\phi} r^2\frac{d\phi}{dE}\, dE = n(t - t_o).$$

Evaluating the left side using Eqn. (5.4) gives $E - e\,\sin E$.

Comparing Eqns. (5.3) and (5.5) the simplicity of Kepler's equation might be appreciated. Here we see another example of the power of using the "proper" variables in celestial mechanics.

It should be observed that neither the true anomaly (Eqn. 5.3) nor the eccentric anomaly (Eqn. 5.5) can be expressed as closed form, explicit functions of time. To obtain E from Eqn. (5.5) for a given time requires the inversion of a transcendental equation. The hundreds of papers in the literature of celestial mechanics dedicated to this problem, still appearing at this date (1988), will not be reviewed here. A simple example, showing the basic approach will suffice.

Consider an elliptic motion with eccentricity $e = 0.1$, and for the sake of simplicity let $a = n = 1$. The given "time" in this case is the same as the mean anomaly. The location of the body is to be found for $l = 30°$ or $t = \pi/6$. This requires the solution of the equation

$$E - 0.1 \sin E = \pi/6,$$

where all terms are expressed in radians.

The process of iteration might start with $E_1 = \pi/6 = 30°$. Another way to obtain a starting value is to use the approximation $\sin E \approx E$, giving $E - .1E \approx \pi/6$, from which $E_1 = 33°19'48''$. Note that establishing a good starting value is important and consequently, the literature treats this problem with considerable attention.

Selecting $E_1 = 30°$, we write the above equation as

$$E = \pi/6 + 0.1 \sin E$$

and substitute on the right side for E the selected E_1 value. In this way we compute the next approximation as $E_2 = \pi/6 + 0.1 \sin E_1 = 32°51'53''$. The process of iteration might be continued until the result is obtained with sufficient accuracy. The difference between the fourth and the fifth iterations is in the fifth decimal, and if this accuracy is sufficient for our purpose we conclude that the solution is $E = 33°7'32''$. The radial distance is computed from Eqn. (5.4) giving $r = 0.9163$. The true anomaly is obtained from Eqn. (5.4) as $f = 36°24'$.

The iteration described above can be expressed by the equation

$$E_{i+1} = l + e \sin E_i \, ,$$

for $i = 1,2,...$ and $E_1 = l$.

Note that for a circular orbit the true and eccentric anomalies for this example are 30°. The reason for the larger value obtained for the elliptic orbit is that the motion took place near the perigee where the velocity is higher than the circular velocity, consequently a larger arc was travelled.

The following summary table of the anomalies might assist the reader as a quick reference.

Symbol	Terminology	Description	Radial Distance
f	True anomaly	Angle at focus of the ellipse	$a(1 - e^2)/(1 + e \cos f)$
E	Eccentric anomaly	Angle at center of the ellipse	$a(1 - e \cos E)$
l	Mean anomaly	$n(t - t_o)$	-

Regularization.

This section intends to offer a broader view of celestial mechanics and to connect seemingly unrelated subjects and techniques. The connection will now be made between the classical subjects of anomalies and the modern method of regularization.

Today we consider regularization as one of the powerful techniques which increases the accuracy of orbit computations at collisions or for close approach trajectories. The method consists of introducing new variables which eliminate the singularities from the equations of motion. The appearance of singularities is usually due to the fact that the gravitational force law introduces inverse square terms in the equations of motion. When the distances become small, these terms dominate and the accuracy of the computation might become questionable.

The credit for introducing regularizing variables usually is given to Sundman (1912), who introduced regularization in order to show the existence of solutions of the differential equations of motion. It is interesting to note that such pure mathematical exercises led to everyday practical techniques used today in our applied orbit mechanics. The combination of a mathematical existence proof and increased accuracy of numerical integration of the orbits of space probes represents an important message to promote the cooperation of engineers and mathematicians.

In orbit mechanics, of course, the distances never become zero, since collision (or impact) occurs before $r = 0$, due to the finite size of the bodies involved; nevertheless, the accuracy of the computation is reduced at close approaches, even when the numerical integration process allows the use of smaller time steps. The introduction of properly selected new variables regularizes the equations of motion, and accuracy can be maintained at the price of using transformations. Both the true and eccentric anomalies are such regularizing variables, and the transformation which introduce them are expressed as

$$df = c\,\frac{dt}{r^2}$$

and

$$dE = an\,\frac{dt}{r}\,.$$

The first relation, as we have seen is identical to the conservation of the angular momentum (Eqn. 3.24), and the second follows from Kepler's equation, using Eqn. (5.4) and (5.5).

Considering the above equations which introduce the true and eccentric anomalies as new independent variables, the origin of regularization could be contributed to Kepler. He certainly had no idea of regularization, especially since to him calculus and differential equations were not known. Furthermore, Kepler's interest in the dynamics of the solar system excluded close approaches. Nevertheless, the use of the eccentric anomaly introduced in his equation allows us to increase the accuracy of our numerical integrations. We might consider this a demonstration of a true genius, or once again we might celebrate the power of combination of two fields, that of Kepler's work to describe planetary motions and our efforts to compute accurate Earth-to-Moon trajectories. Ever since Kepler's work, the subject of regularization was popular. Euler used it to study the straight-line motions in the problem of two bodies. Levi-Civita regularized the restricted problem of three bodies in 1903. K. Stumpff (1949) and Herrick (1965) used basically the same idea when they introduced the concept of universal variables.

The above transformations of the independent variable may be generalized to

$$ds = A\,\frac{dt}{r^m}\,,$$

or

$$ds = \frac{dt}{f(r)}\,,$$

where A and m are constants.

A recently introduced anomaly is known as the intermediate anomaly which is given by $m = 3/2$. Note that $m = 1$ gives the eccentric anomaly and $m = 2$ the true anomaly, with proper selection of the constant A. In general, the above transformations of the time introduce new independent variables which are often denoted by s. When the integration of the differential equations of motion is performed using s as the independent variable, the method is often called "s-integration."

The above transformations, from a numerical point of view, might be considered analytical step regulations since as the distance (r) decreases and close approaches or collisions occur, the time step is to be reduced. Considering the above transformation equations, this means that the integration step using t, will be reduced but the step size using s, might be kept constant. The transformation regulates automatically and analytically the step size.

When transformations of the independent variable are combined with the proper transformations of the dependent variables and the conservation principles (energy and momentum) are used, the result is not only a regularized but also a linearized system of equations describing the gravitational problem. As we have seen in Chapter 4, when the true anomaly is introduced as the new independent variable and $1/r = u$ is used as the new dependent variable, the differential equation of motion became a second order linear differential equation, representing a harmonic oscillator (Eqn. 4.2). Because of the practical numerical importance of such "smoothing transformations," there is a large number of contributions in the literature of modern celestial mechanics dedicated to this subject.

The use of regularization in space dynamics becomes mandatory at close approaches occurring at departures, arrivals, gravity-assist orbit maneuvers (see Chapter 6) and in general, when orbit computations with high accuracy are required. In celestial mechanics the accurate computation of cometary orbits also calls for regularization.

A modern comprehensive treatment of regularization and of the associated linearization is given in the book by Stiefel and Scheifele (1971).

EXAMPLES

1. Compute the semi-major axis of Mars' orbit in astronomical units and in miles, using for its orbital period 1.9 years.

    If the approximate formula (5.2),

    $$\frac{T_1^2}{T_2^2} = \frac{a_1^3}{a_2^3}$$

    is applied to the Earth and Mars, we have

    $$a_M = a_E \left[\frac{T_M}{T_E} \right]^{2/3},$$

    where $a_E = 1, T_E = 1 \ year$ and $T_M = 1.9 \ years$. This gives

$$a_M = 1.9^{2/3} = 1.534 \text{ A.U.} = 142.59 \times 10^6 \text{ miles.}$$

2. If the semi-major axis of a planet from the Sun is 2870×10^6 km, what is its orbital period in years?

First we find the semi-major axis in astronomical units so that with the help of the table of physical constants given in the Appendix we can identify the planet.

$$a = 2870 \times 10^6 \text{ km} = 2870 \times 10^6/(1.496 \times 10^8) = 19.18 \text{ A.U.}$$

This number corresponds to the planet Uranus. Its orbital period becomes

$$T_U = (19.18)^{3/2} = 84 \text{ years.}$$

3. Using the basic data for Explorer 7 as given in Example 2 of Chapter 4, compute the mass of the Earth in kg.

The relation between the period and the semi-major axis is

$$T = 2\pi \left[\frac{a^3}{\mu} \right]^{\frac{1}{2}},$$

where the symbol a in our case stands for the semi-major axis of the satellite $a = 7190.86 \text{ km}$, $\mu = GM_E$, and $T = 1.684 \text{ hours}$. From the above equation we have

$$M_E = \frac{4\pi^2 a^3}{GT^2} = 5.986 \times 10^{24} \text{ kg.}$$

This result shows that the determination of planetary masses is not a simple matter and it is strongly influenced by the observational accuracy.

4. Show that the average value of the distance between the focus and the ellipse is the semi-major axis, provided the averaging is performed with respect to the eccentric anomaly.

Averaging with respect to the eccentric anomaly we have

$$r_{ave} = \frac{1}{2\pi} \int_0^{2\pi} r dE.$$

The relation between r and E is

$$r = a(1 - e \cos E).$$

After substitution and integration, we have

$$r_{ave} = a.$$

If we average with respect to the true anomaly we have

$$r_{ave} = \frac{1}{2\pi} \int_0^{2\pi} r\,df = \frac{p}{2\pi} \int_0^{2\pi} \frac{df}{1 + e\,\cos f}.$$

This integral becomes

$$\int \frac{df}{1 + e\,\cos f} = \frac{2}{\sqrt{1 - e^2}}\, \arctan \left[\frac{1 - e}{1 + e}\right]^{1/2} \tan \frac{f}{2}$$

or using

$$\tan \frac{f}{2} = \left[\frac{1 + e}{1 - e}\right]^{1/2} \tan \frac{E}{2}$$

we have

$$r_{ave} = \frac{p}{2\pi} \frac{2}{\sqrt{1 - e^2}} \left[\frac{E}{2}\right]_0^{2\pi} = \frac{p}{\sqrt{1 - e^2}} = b.$$

It is of interest to find that the time average, or the average with respect to the mean anomaly of r is neither a nor b but

$$r_{ave} = \frac{1}{T} \int_0^T r\,dt = a(1 + e^2/2).$$

This result may be obtained by differentiating Kepler's equation:

$$dt = \frac{dt}{dE}\, dE = (1 - e\,\cos E)\,\frac{dE}{n}$$

and using $r = a(1 - e\,\cos E)$ under the integral sign.

As a conclusion we observe that the mean distance between the focus and the elliptic orbit, in general, is not the length of the semi-major axis. For detailed discussion see R.A. Serafin's article in Celestial Mechanics, Vol. 21, p. 351, 1980 and Taff's (1985) book (see Appendix).

PROBLEMS

1. Compute the average value of the orbital velocity for elliptic motion. Perform the averaging with respect to the eccentric anomaly, true anomaly and mean anomaly.

2. Show that for small values of eccentricity, when e^3 and higher powers can be neglected, the solution of Kepler's equation

becomes

$$E = l + e(1 + e\ cosl)\ sinl.$$

3. Find the values of the eccentric and true anomalies when $l = \pi/4$, $\pi/2$, $3\pi/4$, and π for an elliptic orbit with eccentricity ($e = 0.2$).

4. Show that $\dot{f} = c/r^2$, $\dot{E} = an/r$, and $\dot{l} = n$ on elliptic orbits.

Chapter 6. Orbit Maneuvers

The three problems which might be treated without much difficulty are circularization, escape and transfer orbits.

Circularization.

Let us assume that a satellite is not precisely on a circular orbit. This happens quite frequently when a satellite is placed in orbit, due to some malfunctions of the propulsion units. Then the problem is how to circularize the orbit which is elliptic with given values and a and e. The satellite velocity at perigee is

$$v_P = \left[\frac{\mu}{a} \frac{(1+e)}{(1-e)} \right]^{1/2} ,$$

and if we wish to circularize it at perigee altitude or at a distance,

$$r_P = a(1-e) ,$$

we must change its velocity to

$$(v_c)_P = \left[\frac{\mu}{r_P} \right]^{1/2} = \left[\frac{\mu}{a(1-e)} \right]^{1/2} .$$

In this way the new orbit will be circular at r_P distance from the center of the Earth. The velocity denoted by v_P is larger than $(v_c)_P$ which is expected since it takes more energy, i.e., higher velocity, to go to apogee than to maintain a circular orbit. This expectation may be verified analytically by writing

$$v_P > (v_c)_P .$$

After substitution we have

$$\left[\frac{\mu}{a} \frac{(1+e)}{(1-e)} \right]^{1/2} > \left[\frac{\mu}{a(1-e)} \right]^{1/2} ,$$

or

$$(1+e)^{1/2} > 1 ,$$

which inequality is correct since $e > 0$.

If we wish to circularize the orbit at perigee we must reduce the perigee velocity to the circular velocity. The velocity change is negative, indicating a decrease of the velocity:

$$\Delta v_P = (v_c)_P - v_P < 0$$

or

$$\Delta v_P = \left[\frac{\mu}{a(1-e)} \right]^{1/2} \left[1 - (1+e)^{1/2} \right] . \tag{6.1}$$

Note that the satellite will have to be slowed down, therefore Δv_P will have to be applied in the opposite direction from the motion (retro rockets).

If we wish to circularize the orbit at apogee, we have to increase the elliptic velocity at apogee. This velocity is

$$v_A = \left[\frac{\mu}{a} \frac{(1-e)}{(1+e)} \right]^{1/2}$$

and the circular velocity at this elevation is

$$(v_c)_A = \left[\frac{\mu}{r_A} \right]^{1/2} = \left[\frac{\mu}{a(1+e)} \right]^{1/2} .$$

The elliptic velocity now is smaller than the circular velocity,

$$v_a < (v_c)_A ,$$

since

$$\left[\frac{\mu}{a} \frac{(1-e)}{(1+e)} \right]^{1/2} < \left[\frac{\mu}{a} \frac{1}{(1+e)} \right]^{1/2}$$

or

$$(1-e)^{1/2} < 1 .$$

The velocity increment necessary to circularize is

$$\Delta v_A = (v_c)_A - v_A ,$$

or

$$\Delta v_A = \left[\frac{\mu}{a(1+e)} \right]^{1/2} \left[1 - \sqrt{(1-e)} \right] . \tag{6.2}$$

The increased velocity at apogee will result in a circular orbit of radius $r_A = a(1+e)$, and the reduced velocity at perigee will give a circular orbit with radius $r_P = a(1-e) < r_A$.

If the purpose of the maneuver is to circularize either at apogee or at perigee, we might inquire as to which of these two possible circularizing maneuvers needs less propellant mass expended. This is related to the magnitude of the velocity change or "delta v." This question is formulated analytically by comparing the quantities Δv_P and Δv_A given by Eqns. (6.1) and (6.2).

It can be shown that

$$\Delta v_P > \Delta v_A \ .$$

By substituting from Eqns. (6.1) and (6.2) we obtain

$$\left[\frac{\mu}{a}\right]^{1/2} \frac{\sqrt{1+e}-1}{\sqrt{1-e}} > \left[\frac{\mu}{a}\right]^{1/2} \frac{1-\sqrt{1-e}}{\sqrt{1+e}} \ ,$$

or

$$1 + e - \sqrt{1+e} > \sqrt{1-e} - (1-e) \ ,$$

or

$$2 > \sqrt{1+e} + \sqrt{1-e} \ .$$

Squaring both sides gives, after cancellation,

$$1 > \sqrt{1-e^2} \ ,$$

which concludes the proof.

Circularization, therefore, should be performed at apogee (see Figure 6.1).

Escape.

The second problem to be discussed is related to escape. By this we mean that we have a satellite or space probe on an elliptic orbit around the Earth and we wish to send it on an interplanetary mission, or for some other reason, we wish to change its orbit from elliptic to a parabolic escape orbit. The question is if the velocity change should be executed at perigee or apogee. Both maneuvers require an increase in the velocity from v_P or from v_A to v_e which is the escape velocity. The location of the optimal position where the maneuver is to be performed is not obvious for the following reason. At perigee the vehicle's velocity is higher than at apogee so we might expect that the required Δv is smaller at perigee than at apogee. On the other hand the perigee is closer to the Earth than the apogee, so escape might occur easier at apogee, which is farther from the center of gravitational attraction than at perigee where the attraction is stronger.

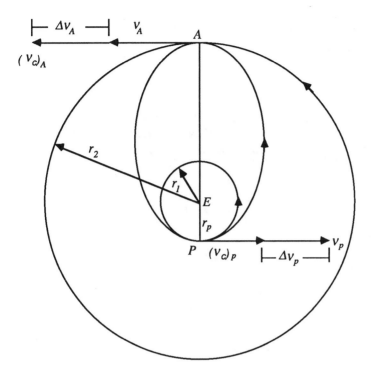

Figure 6.1. Circularization Maneuvers.

First the formula for escape velocity will be derived by considering the energy conservation as given by Eqn. (3.19) and by the discussion following Eqn. (4.9). For parabolic escape orbits $k = 0$, and the energy equation gives

$$v_e = \left[\frac{2\mu}{r} \right]^{1/2}, \tag{6.3}$$

or

$$v_e = \sqrt{2}\, v_c \;.$$

Note that hyperbolic escape orbits have higher velocities, but at this point we are interested in the minimal velocity at which escape occurs.

The above formula for escape velocity may be obtained from the energy equation

$$v^2 = \frac{2\mu}{r} + k$$

considering elliptic, parabolic or hyperbolic orbits.

For elliptic orbits

$$v^2 = \mu\left[\frac{2}{r} - \frac{1}{a}\right] \, ,$$

and escape is associated with an orbit for which $a \to \infty$. Therefore $v_e = \sqrt{2\mu/r}$. For parabolic orbits the above derivation of Eqn. (6.3) offers the desired result. Finally, for hyperbolic orbits the equation of energy conservation gives, with $k = 1/a$,

$$v^2 = \mu\left[\frac{2}{r} + \frac{1}{a}\right] \, ,$$

see Eqn. (7.2).

If, as $r \to \infty$ we wish to have $v = 0$, we must once again have $1/a = 0$ and $v_e = \sqrt{2\mu/r}$. In general as $r \to \infty$, $v \to \sqrt{\mu/a}$, so the velocity asymptotically approaches the value $\sqrt{\mu/a}$ on a hyperbolic escape orbit, i.e. at infinity the velocity, known as "hyperbolic excess velocity," is not zero.

Now we return to the optimization of escape from an elliptic orbit. The velocity required for escape at perigee is

$$(v_e)_P = \left[\frac{2\mu}{a(1-e)}\right]^{1/2}$$

and at apogee

$$(v_e)_A = \left[\frac{2\mu}{a(1+e)}\right]^{1/2} .$$

We see that

$$(v_e)_P > (v_e)_A$$

since at perigee the vehicle is closer to the center of attraction than at apogee.

The perigee and apogee velocities on an elliptic orbit are given by Eqns. (4.13). These velocities of course are smaller than the corresponding escape velocities, as expected. This can be seen from the inequalities,

$$(v_e)_P = \sqrt{2}\left[\frac{\mu}{a(1-e)}\right]^{1/2} > \sqrt{(1+e)}\left[\frac{\mu}{a(1-e)}\right]^{1/2} = v_P$$

and

$$(v_e)_A = \sqrt{2}\left[\frac{\mu}{a(1+e)}\right]^{1/2} > \sqrt{1-e}\left[\frac{\mu}{a(1+e)}\right]^{1/2} = v_A.$$

The first of these inequalities can be reduced to $2 > 1+e$ or $1 > e$, and the second to $1 > -e$.

The velocity increment needed for escape at perigee is

$$\Delta v_P = (v_e)_P - v_P$$

and at apogee is

$$\Delta v_A = (v_e)_A - v_A.$$

After substitution we obtain

$$\Delta v_P = \left[\frac{\mu}{a(1-e)}\right]^{1/2}\left[\sqrt{2} - \sqrt{1+e}\right]$$

and

$$\Delta v_A = \left[\frac{\mu}{a(1+e)}\right]^{1/2}\left[\sqrt{2} - \sqrt{1-e}\right].$$

In order to show that $\Delta v_A > \Delta v_P$ we consider the inequality

$$\frac{\sqrt{2}-\sqrt{1-e}}{\sqrt{1+e}} > \frac{\sqrt{2}-\sqrt{1+e}}{\sqrt{1-e}},$$

which can be reduced to $1 > e$.

We conclude that escape is to be executed at perigee if velocity differences are to be optimized.

Hohmann's Transfer Orbit.

The third idea concerning orbit changes is known as Hohmann's (1925) tangential elliptic transfer orbit which optimizes the total velocity change required when we wish to change from one circular orbit to another concentric circular orbit. Figure 6.1 once again describes the operation, but the physical meaning of the figure is slightly changed. The two circular orbits have the radii, $r_1 = r_P = \overline{EP}$ and $r_2 = \overline{EA}$.

Consider elliptic transfer orbit from a parking orbit with radius r_1 to another (larger) circular orbit with radius r_2.

The semi-major axis of the ellipse is $a = (r_1 + r_2) / 2$, the perigee distance is $r_P = r_1 = a(1 - e)$, and the apogee distance is $r_A = r_2 = a(1 + e)$. For given values of r_1 and r_2 the eccentricity may be computed from the above relations. We have

$$e = \frac{r_2 - r_1}{r_1 + r_2}.$$

The perigee and apogee velocities of the elliptic orbit may be computed from the energy equation, giving

$$v_P^2 = \mu\left[\frac{2}{r_1} - \frac{1}{a}\right]$$

and

$$v_A^2 = \mu\left[\frac{2}{r_2} - \frac{1}{a}\right],$$

or

$$v_P = \sqrt{2\mu}\left[\frac{r_2}{r_1}\right]^{1/2}\frac{1}{\sqrt{r_1 + r_2}}$$

and

$$v_A = \sqrt{2\mu}\left[\frac{r_1}{r_2}\right]^{1/2}\frac{1}{\sqrt{r_1 + r_2}}.$$

The velocities at P and A on the circular orbits with radii r_1 and r_2 are

$$(v_c)_P = \left[\frac{\mu}{r_1}\right]^{1/2} \text{ and } (v_c)_A = \left[\frac{\mu}{r_2}\right]^{1/2}.$$

Since $(v_c)_P < v_P$ and $(v_c)_A > v_A$, the transfer orbit requires an increase in the velocity from $(v_c)_P$ to v_P at perigee and another increase in velocity from v_A to $(v_c)_A$ at apogee. This is expected since at P the circular velocity must be increased in order to reach a higher elevation (larger distance, r_2) and at A the elliptic apogee velocity (v_A) must be increased to obtain a circular orbit with radius r_2. If this second velocity increase would not be executed, the vehicle would stay on its elliptic orbit, and it would return to P.

In addition to the physical reasoning we might also show analytically the correctness of the above inequalities by substituting the equations given for $(v_c)_P$, $(v_c)_A$, v_P and v_A. After simplifications, both inequalities lead to

$r_2 > r_1$.

The velocity increase at P is

$$(\Delta v)_P = v_P - (v_c)_P ,$$

and at A we have

$$(\Delta v)_A = (v_c)_A - v_A .$$

Substituting the previously given results we have for the total velocity change

$$\Delta v = (\Delta v)_P + (\Delta v)_A ,$$

or

$$\Delta v = \sqrt{2\mu}\left\{ \frac{1}{\sqrt{r_1}}\left[\left(\frac{r_2}{r_1 + r_2} \right)^{1/2} - 1 \right] - \frac{1}{\sqrt{r_2}}\left[\left(\frac{r_1}{r_1 + r_2} \right)^{1/2} - 1 \right] \right\}.$$

The transfer time is one half of the period of the elliptic orbit. This period might be obtained from Kepler's law and the transfer time becomes

$$T/2 = \pi\left[\frac{(r_1 + r_2)^3}{8\mu} \right]^{1/2}.$$

This chapter is closed with a remark concerning the origin of the changes in velocities which are required for the orbit maneuvers discussed. These changes can come from propulsion or from more natural sources. The second type of orbit changes are usually referred to as sling-shot, gravity-assist or fly-by maneuvers. The probe makes a close approach trajectory to a natural satellite or planet and its velocity is increased or decreased, depending on the geometry of its orbit. Examples are Mariner 10 with Venus and Pioneer 10 and 11 with Jupiter.

Out-of-plane orbital changes are discussed at the end of Chapter 9, in Example 2.

W. Hohmann's original work was published in his book, *Die Erreichbarkeit der Himmelkörper*, Oldenburg Publ., Munich (1925). (English translation published by NASA, Washington, D.C., 1960). Note that the exact proof of Hohmann's result appeared only in 1961 showing that the optimum condition is limited by the requirement that $r_2/r_1 < 11.94$. See Ehricke's Vol. 2, Chapter 6 (1962) and Battin's Sections 9.2 and 11.3 (1987), mentioned in the Appendix. An easy to read report with engineering orientation is by J. B. Eades, "Orbital Transfer," Bulletin of the Virginia

Polytechnic Institute, Blacksburg, Virginia, 1965.

EXAMPLES

1. What velocity increment would be needed to have the Earth escape from the Sun?

    The average velocity of the Earth around the Sun is given by

    $$v_E = \left[\frac{G(m_S + m_E)}{a_E} \right]^{1/2} = 29.78 \; km/sec.$$

    The escape velocity is

    $$v_e = \sqrt{2} \, v_E = 42.12 \; km/sec.$$

    The required velocity increment is

    $$v_e - v_E = 12.34 \; km/sec.$$

2. If the eccentricity of the Earth's orbit is $e = 0.0167$ what is the ratio of the perihelion and aphelion velocities?

    Using Eqn. (4.14) we have

    $$\frac{v_P}{v_A} = \frac{1.0167}{0.9833} = 1.03397 \,.$$

    Using the approximate formula,

    $$\frac{v_P}{v_A} \approx 1 + 2e \,,$$

    we have 1.0334. (See Chapter 4, following Eqn. 4.14.)

    The difference between the two velocities is obtained from Eqns. (4.13):

    $$v_P - v_A = \left[\frac{\mu}{a} \right]^{1/2} \left[\left[\frac{1+e}{1-e} \right]^{1/2} - \left[\frac{1-e}{1+e} \right]^{1/2} \right]$$

    or

    $$\Delta v = \left[\frac{\mu}{a} \right]^{1/2} \frac{2e}{\sqrt{(1-e^2)}} \,.$$

    The numerical value is $0.0334 \times 29.8 = 0.995 \; km/sec$.

Note that an approximate equation for the velocity difference, valid for small values of the eccentricity is

$$\Delta v \approx 2e \left[\frac{\mu}{a} \right]^{1/2},$$

which in our case gives the same result, i.e. $\Delta v = 0.995 \; km/sec$.

3. The escape velocity from a planet is computed from Eqn. (6.3):

$$v_e = \left[\frac{2\mu}{r} \right]^{\frac{1}{2}},$$

where r is the distance from the center of the planet to the escaping body and $\mu = GM_P$. The escape velocities from the surfaces of planets become

Planet	Mercury	Venus	Earth	Mars	Jupiter	Saturn	Uranus	Neptune	Pluto
Escape Velocity (km/sec)	4.25	10.36	11.18	5.024	59.57	35.56	21.33	23.76	1.16

Note that planetary atmospheres will influence the above-computed escape velocities.

The escape velocity from the surface of the Sun is 617.53 km/sec and from the Moon is 2.375 km/sec.

The escape velocity from the surface of the Earth is 11.18 km/sec. Throwing a stone at a velocity of 1.118 km/sec will not escape unless we are on a planet which is smaller than the Earth. The formula for the velocity of escape is

$$v_e = \left[\frac{2GM_P}{R_P} \right]^{1/2} = R_P \left[\frac{8}{3} \pi G \rho_P \right]^{1/2}$$

where ρ_P is the density of the planet. This new formula is obtained by writing $(4\pi/3)R_P^3 \rho_P$ for the mass of the planet. Since the mean density of the Earth is 5.52 g/cm^3, the radius of a planet (with the Earth's density) should be $R_P = R_E/10 = 637.8 \; km$ to have 1.118 km/sec as the escape velocity from the surface.

Note that the escape velocity is proportional to the planetary radius, as long as the density and the gravitational constant are not altered.

4.   In this example a Hohmann transfer orbit is established from a circular parking orbit to a geosynchronous orbit (see Example 2, Chapter 1).

  If the velocity is increased tangentially on the circular parking orbit, the point where this occurs will be the perigee of the transfer orbit. The circular velocity of a satellite at elevation $h = 300 \ km$ is

$$v_c = \left[\frac{\mu}{R_E + h} \right]^{1/2} = 7.726 \ km/sec.$$

  The elliptic orbit's apogee distance from the Earth's center is $d = 42240.14 \ km$, and the semi-major axis of the transfer orbit is

$$a = \frac{h + R_E + d}{2} = 24459.14 \ km.$$

  The eccentricity of the transfer orbit is

$$e = \frac{d - R_E - h}{d + R_e + h} = 0.727 \ ,$$

which follows from the expression for the perigee distance,

$$r_P = a(1 - e) \ ,$$

where $r_P = h + R_e$, and $a = (h + R_e + d)/2$.

  The perigee and apogee velocities are

$$v_P = \left[\frac{\mu}{a} \right]^{1/2} \left[\frac{1 + e}{1 - e} \right]^{1/2} = \left[\frac{2\mu d}{(d + R_e + h)(R_e + h)} \right]^{1/2} =$$

$$10.153 \ km/sec$$

and

$$v_A = \left[\frac{\mu}{a} \right]^{1/2} \left[\frac{1 - e}{1 + e} \right]^{1/2} = \left[\frac{2\mu(R_E + h)}{d(d + R_E + h)} \right]^{1/2} = 1.605 \ km/sec.$$

  The radial distance of the original circular parking orbit from the center of the Earth is $R_e + h = 6678.14 \ km$, and the original circular velocity is $v_c = 7.726 \ km/sec$.

  The velocity increment at perigee is $\Delta v_P = v_P - v_c = 2.427 \ km/sec$.

  The circular velocity at apogee distance is $v'_c = \sqrt{\mu/d} = 3.072 \ km/sec$, but the velocity on the elliptic orbit at apogee is smaller than this. Therefore, if a circular orbit is to be obtained at apogee, we need a velocity increment of $\Delta v_A = v'_c - v_A = 1.467\ km/sec$.

The total velocity change is $\Delta v_t = \Delta v_P + \Delta v_A = 3.894\ km/sec$.

The transfer time is half of the period of the elliptic orbit:

$$T/2 = \pi \left[\frac{a^3}{\mu} \right]^{1/2} = 5\ hour\ 17\ min\ 14.6\ sec.$$

5. The idea of the elliptic Hohmann transfer orbit can be used to establish a lunar trajectory. Note that the following equations offer only approximations since the Moon's effect on the trajectory is neglected.

The parking orbit is at elevation h, so the circular velocity is

$$v_c = \left[\frac{\mu}{h + R_e} \right]^{1/2}.$$

The semi-major axis of the transfer orbit is

$$a = \frac{R_e + h + d_{EM}}{2},$$

where d_{EM} is the distance between the centers of the Earth and the Moon.

The eccentricity of the transfer orbit is

$$e = \frac{d_{EM} - h - R_e}{d_{EM} + h + R_e}.$$

The perigee velocity is

$$v_P = \left[\frac{2\mu d_{EM}}{(h + R_e)\ (h + R_e + d_{EM})} \right]^{1/2}$$

and the velocity increment on the parking orbit becomes

$$\Delta v_P = v_P - v_c .$$

The arrival velocity at the Moon (neglecting in the first approximation the gravitational effect of the Moon) is the apogee velocity of the elliptic transfer orbit,

$$v_A = v_P \frac{h + R_e}{d_{EM}},$$

which equation utilizes the conservation of the angular momentum. Since $h + R_e < d_{EM}$, $v_A < v_P$ as expected. The circular velocity at apogee is the lunar velocity which is higher than the arrival velocity; therefore, a new boost is required:

$$\Delta v_A = v_M - v_A \, ,$$

where v_M is the Moon's (approximate) circular velocity.

The transfer time is once again half the period of the transfer ellipse:

$$T/2 = \pi \left[\frac{h + R_e + d_{EM})^3}{8\mu} \right]^{1/2} .$$

6. This example inverts the usual problem of orbit changes and intends to establish the new orbit when a certain velocity change occurs. If the velocity increases from circular velocity (v_c) to v, then the change is $\Delta v = v - v_c$ or $v = v_c(1 + \Delta v/v_c)$. The dimensionless velocity change is $x = \Delta v/v_c$ which, when multiplied by 100, represents the percentage change of the velocity.

  If the circular velocity is increased, it becomes the perigee velocity or

$$v_P = v_c(1 + x) .$$

The circular velocity at perigee distance was

$$v_c = \left[\frac{\mu}{a(1-e)} \right]^{1/2} ,$$

and the increased perigee velocity is

$$v_P = \left[\frac{\mu}{a} \frac{(1+e)}{(1-e)} \right]^{1/2} .$$

  After the substitution of the expressions for v_c and v_P into the relation $v_P = v_c(1 + x)$, we have

$$\sqrt{1 + e} = 1 + x \, ,$$

or

$$e = x(2 + x) \approx 2x \, ,$$

where the approximation is valid for small x and small e values.

  A 10 percent velocity increase on the circular orbit results in an elliptic orbit with eccentricity $e = 0.1(2 + 0.1) = 0.21$.

  The percentage velocity increase required for escape can be obtained when the relation $\sqrt{1 + e} = 1 + x$ is used with $e = 1$. The result is $x = \sqrt{2} - 1$, i.e. 41.42 percent increase (or more) of the circular velocity is required for escape.

The orbital parameters of the elliptic orbit (when $x < 0.4142$) can be expressed as functions of x. For instance if the altitude of the original circular orbit is h, the semi-major axis of the elliptic orbit becomes

$$a = \frac{h + R_e}{1 - 2x - x^2} .$$

For a 10 percent velocity increase, the semi-major axis becomes $a = 1.266(h + R_e)$.

The apogee velocity becomes

$$v_A = \frac{1 - e}{1 + e} v_P ,$$

or

$$v_A = \frac{1 - x(2 + x)}{(1 + x)^2} v_P ,$$

or

$$v_A = \frac{1 - x(2 + x)}{1 + x} v_c .$$

For 10 percent velocity increase the apogee velocity becomes $0.7182 v_c$.

7. The orbit of the artificial satellite Explorer 6, known as 1959 δ2 has a perigee distance of $r_P = 6622.6 \, km$ and an apogee distance of $r_A = 48201 \, km$. (Note that these are not heights but distances from the center of the Earth to a point on the orbit.)

    The semi-major axis is

$$a = \frac{1}{2}(r_P + r_A) = 27411.8 \, km.$$

    The eccentricity is computed by using the equations for the perigee and apogee distances:

$$r_P = a(1 - e) \quad \text{and} \quad r_A = a(1 + e) .$$

    From these we have

$$e = \frac{r_A - r_P}{r_A + r_P} = 0.758.$$

The semi-minor axis and the semi-latus rectum are

$$b = a\sqrt{1 - e^2} = 17879 \ km$$

and

$$p = a(1 - e^2) = 11662 \ km.$$

The mean motion is computed from Kepler's law:

$$GM_E = n^2 a^3.$$

From this we have $n = 1.4 \times 10^{-4} \ rad/sec = 12.1 \ rad/day$.

The orbital period of the satellite is

$$P = \frac{2\pi}{n} = 12.5 \ hours.$$

The perigee and apogee velocities are

$$v_P = \left[\frac{\mu}{a} \right]^{1/2} \left[\frac{1 + e}{1 - e} \right]^{1/2} = 10.28 \ km/sec$$

and

$$v_A = \left[\frac{\mu}{a} \right]^{1/2} \left[\frac{1 - e}{1 + e} \right]^{1/2} = 1.42 \ km/sec.$$

Note that once the perigee velocity is known, the apogee velocity can be computed by

$$v_A = v_P \frac{1 - e}{1 + e}.$$

The advantage of using this method is the simplicity of the formula. The disadvantage is that if the computed value of v_P is in error, v_A will be also in error.

If this satellite is at perigee when we start observing time (i.e. at $t = 0$, $f = l = E = 0$; see the table of anomalies given in Chapter 5), its location at $t = 62.5 \ min$. as it moves counter clockwise can be established as follows.

The mean anomaly at this time is $l = nt = 0.526 \ rad = 30°$.

The eccentric anomaly can be computed from

$$l = E - e \ sinE$$

by iteration, starting with $E_o = 30°$. The solution is $E \approx 71.27 \ degrees$.

The true anomaly is computed from

$$\tan\frac{f}{2} = \left[\frac{1+e}{1-e}\right]^{1/2} \tan\frac{E}{2} \, ,$$

giving $f = 125.271 \ degrees$.

The distance from the center of the Earth can be computed either from $r = a \ (1 - e \ \cos E)$ or from $r = p/(1 + e \ \cos f)$.

The results are 20739.75 km and 20739.93 km showing an error of 0.177 km or less than 0.001 percent. This error is due to the error of the approximate solution obtained by solving Kepler's transcendental equation by iteration.

In order to circularize the orbit at apogee a velocity boost is required. This circular velocity is

$$(v_A)_c = \left[\frac{\mu}{r_A}\right]^{1/2} = \left[\frac{\mu}{a}\right]^{1/2} \frac{1}{\sqrt{1+e}} = 2.876 \ km/sec \, ,$$

and the velocity increment is

$$(\Delta v_A)_c = 2.876 - 1.42 = 1.456 \ km/sec.$$

To circularize at the perigee the satellite will have to slow down. The circular velocity at perigee altitude is

$$(v_P)_c = \left[\frac{\mu}{r_P}\right]^{1/2} = \left[\frac{\mu}{a}\right]^{1/2} \frac{1}{\sqrt{1-e}} = 7.75 \ km/sec \, ,$$

and the velocity change is

$$(\Delta v_P)_c = 10.28 - 7.75 = 2.53 \ km/sec.$$

As expected, the required velocity change is larger at perigee than at apogee; therefore, circularization is to be performed at apogee.

To escape at apogee the velocity must be

$$(v_A)_e = \sqrt{2}(v_A)_c = 4.07 \ km/sec.$$

The increase of velocity is

$$(\Delta v_A)_e = 4.07 - 1.42 = 2.65 \ km/sec.$$

The escape velocity at perigee is

$$(v_P)_e = \sqrt{2}(v_P)_c = 10.96 \ km/sec.$$

The increase of velocity is

$$(\Delta v_P)_e = 10.96 - 10.28 = 0.68 \ km/sec.$$

Note that $(\Delta v_A)_e > (\Delta v_P)_e$ and, therefore, it is more efficient to execute the escape maneuver at perigee than at apogee.

PROBLEMS

1. A space vehicle approaching a planet on a hyperbolic orbit (relative to the planet) wishes to be captured. Find the change (reduction) of the (hyperbolic) velocity at the pericenter in order to obtain an elliptic orbit with eccentricity e. The hyperbolic excess velocity of the probe is v_∞ and the distance of the pericenter is r_p.

2. A space probe is orbiting a planet on an elliptic orbit with apocenter r_A and pericenter r_P. Find the change of the pericenter if the velocity at the apocenter is increased by Δv_A.

Chapter 7. Hyperbolic and Parabolic Orbits

The reader will find applications of this chapter in the computation of high speed trajectories of space probes and missiles as well as in cometary dynamics.

In the equation of energy (3.19) the constant k represents the total energy. The relation of this constant to the eccentricity of the orbit is given by Eqn (4.9),

$$e = \left[1 + \frac{kc^2}{\mu^2} \right]^{1/2} .$$

From this we conclude that for positive total energy, the orbit is a hyperbola and for zero total energy, the orbit is a parabola. To assist our physical intuition we write the energy equation in the form

$$k = v^2 - \frac{2\mu}{r} ,$$

from which we observe at points close to the center of attraction (small values of r), we require high velocities in order to have hyperbolic or parabolic orbits. In the following, hyperbolic orbits will be treated in more detail since they occur more frequently than parabolic orbits.

The polar equation of a hyperbola is the same as for an ellipse,

$$r = \frac{p}{1 + e \, \cos f} . \tag{7.1}$$

The semi-latus rectum and the energy constant are given by

$$p = a(e^2 - 1), \text{ and by } k = \frac{\mu}{a} . \tag{7.2}$$

The relation between the constant of energy k and the semi-major axis of the ellipse is $k = -\mu/a$, and of the hyperbola is $k = \mu/a$. The above definitions of p and a result in positive values for these geometric parameters. Note that this is not uniform in the literature of celestial mechanics.

To gain some experience with the similarities and differences between elliptic and hyperbolic orbits we show the boundedness of elliptic orbits and the unboundedness of the hyperbolic orbits.

Since $1 + e \cos f$ is not zero for any value f when $0 < e < 1$, the orbit of an ellipse is bounded. Analytically we may write

$$r_P \leq r \leq r_A \,,$$

where $r_A = a(1 - e)$ and $r_P = a(1 + e)$ represent the apocenter and the pericenter distances. From the above inequalities we have

$$a(1 - e) \leq \frac{a(1 - e^2)}{1 + e \cos f} \leq a(1 + e) \,,$$

or

$$\frac{1}{1 + e} \leq \frac{1}{1 + e \cos f} \leq \frac{1}{1 - e} \,.$$

These inequalities are obtained by dividing the previous line by the positive quantity $(1 - e^2)$. Taking now the reciprocals (and changing the directions of the inequalities) we have

$$1 + e \geq 1 + e \cos f \geq 1 - e \,.$$

Subtracting 1 and dividing by e, we have

$$1 \geq \cos f \geq -1 \,.$$

We conclude the well-known fact that the radial distance r is bounded between the perigee and the apogee distances, since the true anomaly f varies from 0 to 2π. For a hyperbolic orbit the geometry is slightly more complicated since the quantity $1 + e \cos f$ is not limited from below by $(1 - e)$, and it may become zero when $\cos f \to -1/e$. This allows $r \to \infty$. The geometrical description of hyperbolic orbits is shown in Figure 7.1.

The left branch of the hyperbola represents the orbit governed by the central body located at F. The center of the hyperbola is at O and the pericenter at P. The distance between the center and the pericenter is a which is the semi-major axis. Note the analogy with elliptic orbits as shown in Figures 4.1 and 7.1. The semi-major axis is $a = \overline{PO}$ and the distance $\overline{OF} = ae$, for both kinds of orbits. The pericenter distance for a hyperbola is $\overline{FP} = a(e - 1)$ but for an ellipse it is $a(1 - e)$. The semi-minor axis is $b = a \sqrt{e^2 - 1}$. Note that for an ellipse the sum of the distances from the foci is $2a$ and for a hyperbola the difference between those distances is $2a$. For instance the distance $\overline{F'P}$ is $ae + a = a(e + 1)$ and the distance $\overline{FP}$ is $a(e - 1)$. The difference is $2a$. Similarly the length of the semi-latus rectum is $\overline{FQ} = p$.

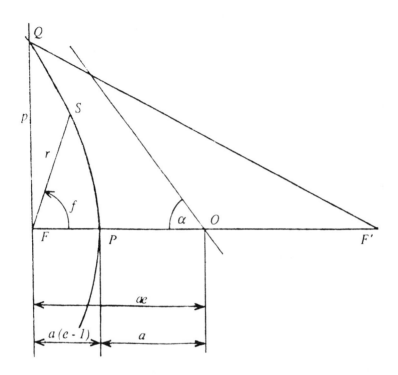

Figure 7.1. Hyperbolic Orbit.

The distance of the point Q from F' is $\overline{QF'} = \sqrt{p^2 + (2ae)^2}$. In order to verify the previously given formula for p we write

$$\overline{QF'} - \overline{QF} = 2a \ ,$$

or

$$\sqrt{p^2 + 4e^2a^2} - p = 2a \ ,$$

from which $p = a(e^2 - 1)$.

The angle of the asymptotes may be obtained from finding the angle f_o from the equation $1 + e \cos f_o = 0$, since this corresponds to $r \to \infty$. The solution is

$$f_o = \arccos(-1/e)$$

and therefore $\alpha = \pi - f_o$ or $\alpha = \arccos(1/e)$. Note that this angle may also be obtained as $\alpha = \arctan(b/a) = \arctan\sqrt{e^2 - 1}$.

High eccentricity ellipses have small semi-major axes because of the relation $b = a\sqrt{1 - e^2}$, and the orbit becomes a circle when $e = 0$. For hyperbolas, large eccentricity means large angles of asymptotes and large values of p and b. When the value of e is small for a hyperbola (i.e. it is slightly larger than 1), $\alpha \approx 0$.

The equation of an ellipse using rectangular coordinates with origin at the center of the ellipse and the major axis along the x axis is

$$\frac{x^2}{a^2} + \frac{y^2}{b^2} = 1 \ ,$$

which equation may be written in parametric form using the eccentric anomaly as

$$x = a \ \cos E \ ,$$

$$y = b \ \sin E \ .$$

The corresponding equations for a hyperbola are

$$\frac{x^2}{a^2} - \frac{y^2}{b^2} = 1$$

and

$$x = a \ \cosh F \ ,$$

$$y = b \sinh F \ ,$$

where F is the hyperbolic eccentricity anomaly.

The hyperbolic functions are defined by their relations to exponential functions as follows

$$\cosh F = \frac{e^F + e^{-F}}{2} \ ,$$

and

$$\sinh F = \frac{e^F - e^{-F}}{2} \ ,$$

which relations correspond to Euler's formulas for trigonometric functions. Note that in the above definitions the symbol e is not the eccentricity but represents $e = 2.71828 \ ...$, the base of natural logarithm. The inverse functions of the trigonometric functions are sometimes called the "arcs" such as arcsin$x$, and sometimes are denoted by $\sin^{-1}x$. The corresponding inverse functions of the hyperbolic functions are known as the "areas" which, because of the exponential relations given above, might be expressed as logarithms. The following relations are standard formulas which can be located in any book on functions, and they are given here since they are used when time durations on hyperbolic orbits are to be evaluated.

The inverse trigonometric functions of $u = \sin v$, $u = \cos v$ and $u = \tan v$ are $v = \arcsin u$, $v = \arccos u$ and $v = \arctan u$.

The corresponding hyperbolic functions of $u = \sinh v$, $u = \cosh v$, and $u = \tanh v$ are $v = \ln(u + \sqrt{u^2 + 1})$, $v = \ln(u \pm \sqrt{u^2 - 1})$, and $v = 1/2 \ln\dfrac{1 + u}{1 - u}$.

The relation between the hyperbolic eccentric anomaly F and the true anomaly f (corresponding to Eqn. 6.4) is given by

$$r = \frac{a(e^2 - 1)}{1 + e \cos f} = a(e \cosh F - 1) , \qquad (7.3)$$

which may be obtained when the projections of the point on the hyperbola on the x and y axes are evaluated using

$$r \cos f = a(e - a \cosh F)$$

and

$$r \sin f = a\sqrt{e^2 - 1} \sinh F .$$

Another form of the relation between f and F is

$$\tan(f/2) = \left[\frac{e + 1}{e - 1}\right]^{1/2} \tanh(F/2) , \qquad (7.4)$$

showing similarity to the corresponding relation applicable to elliptic orbits.

As expected, the eccentric anomaly for hyperbolas is related to an area, just as in the case of elliptic motion it is an angle or arc. Consider the areas formed by the lines OP, OS and by the curve representing the hyperbola between P and S. Then the eccentric anomaly is given by

$$F = 2\frac{\text{Area}(POS)}{a^2} .$$

This geometric interpretation of F is given only for completeness' sake since it is seldom used for computational purposes.

Kepler's equation for hyperbolic orbits is

$$n(t - t_o) = e(\sinh F) - F . \qquad (7.5)$$

This equation is obtained from the conservation of angular momentum as before since from

$$r^2\dot{\phi} = c \text{ or } r^2\dot{f} = c$$

we have

$$c\int dt = \int r^2 \frac{df}{dF} dF .$$

Here r^2 is substituted from the second part of Eqn. (7.3) and df/dF is computed by differentiating Eqn. (7.3) on both sides.

At this point we turn to parabolic orbits for which the total energy is zero.

Consider the energy equation in the form given before as Eqn. (3.19):

$$v^2 = \frac{2\mu}{r} + k . \tag{7.6}$$

If the velocity at infinity is zero, the above equation results in $k = 0$.

The polar equation of the parabola is

$$r = \frac{p}{1 + \cos\phi} , \tag{7.7}$$

which is the special case of Eqn. (4.4) discussed before, and it is obtained when the eccentricity $e = 1$. The length of the semi-latus rectum of a parabola is obtained from the pericenter distance since from Eqn. (7.7) we have for $\phi = 0$

$$2\,r_P = p .$$

The value of the constant of the angular momentum is the same as before, $c = \sqrt{\mu p}$ which might be obtained from the following simple exercise.

At perigee the velocity is

$$v_P = \left[2\frac{\mu}{r_P} \right]^{1/2} ,$$

and the perigee distance is $r_P = p/2$. The angular momentum at this point is

$$c = r_P v_P = \left[\frac{2\mu r_P^2}{r_P} \right]^{1/2} = \sqrt{\mu p} .$$

The time dependence on the parabolic orbit might be obtained directly from the conservation of angular momentum, $c = r^2\dot{\phi}$, or

$$c(t - t_o) = \int_0^\phi r^2 d\phi ,$$

or

$$\sqrt{\mu p}\,(t - t_o) = p^2 \int_0^\phi \frac{d\phi}{(1 + \cos\phi)^2} \ ,$$

where $t = t_o$ and $\phi = 0$ correspond to perigee passage. The integral on the right side may be evaluated considering that $(1 + \cos\phi)^2 = 4\,\cos^4(\phi/2)$ and that

$$\int \frac{dx}{\cos^4 x} = \int \frac{\sin^2 x}{\cos^4 x}\,dx + \int \frac{\cos^2 x}{\cos^4 x}\,dx \ ,$$

from which

$$\int \frac{dx}{\cos^4 x} = \frac{1}{3}\tan^3 x + \tan x \ .$$

Consequently, the elapsed time on parabolic orbits becomes

$$\left[\frac{\mu}{p^3}\right]^{1/2} (t - t_o) = \tan\frac{f}{2} + \frac{1}{3}\tan^3\frac{f}{2} \ , \qquad (7.8)$$

where $f = \phi$ is the true anomaly. Note the similarity between the expression for the mean anomaly for elliptic orbits $l = n\,(t - t_o)$ and Eqn. (7.8). For elliptic orbits $n = \sqrt{\mu/a^3}$ and now the semi-latus rectum takes over the role of the semi-major axis. Since the solution of cubic equations is known in explicit form, the time dependence of the true anomaly and consequently the functional dependence of the radial distance on the time can be given in closed form. Eqn. (7.8) is known as Barker's equation and was published by Euler in 1743.

Eqn. (7.8) shows that, as $f \to \pi$, $t \to \infty$. Furthermore the $p = 2\,r_P$ relation for an ellipse is satisfied when $a\,(1 - e^2) = 2\,a\,(1 - e)$ or when $(1 - e^2) = 0$, which gives $e = 1$ as expected. The same limit process applies to hyperbolic orbits.

Note that the term $\tan(f/2)$ can also be used to express the radial distance, and Eqn. (7.7) may be written as

$$r = \frac{p}{2}\left[1 + \tan^2\frac{f}{2}\right] \ .$$

This chapter is concluded with a remark concerning the limit $e \to 1$ for conic sections. Parabolic orbits, besides the $e = 1$ condition, must also satisfy the $a \to \infty$ requirement.

If the length of the semi-major axis is finite and $e = 1$, we have flat (straight-line) elliptic and hyperbolic orbits as mentioned before, since the length of the semi-minor axis and the length of the semi-latus rectum become zero.

Another note of considerable interest is that parabolic orbits display a singular property. By this we mean that all those orbits for which the total energy is negative ($k < 0$) are ellipses and all those orbits for which the energy is positive ($k > 0$) are hyperbolas, but only one special value of the energy ($k = 0$) results in parabolic orbits. This fact has some interesting practical and theoretical consequences. Since the initial conditions are usually not known exactly in practical problems, it seldom happens that the energy constant is exactly zero. The same applies to the case when the nature of the orbit is established by observations, which furnish only approximations for the energy. In other words, instead of having exactly zero, we might have small positive or small negative values for the energy of the actual two-body orbit. The situation becomes even more complicated when the values of the physical constants which enter the energy equation are considered. Since the values of the constants of gravity and of the central mass are known only approximately (within error limits), the computed total energy will be also an approximation. We conclude that for practically important cases the constant of energy is determined only approximately. When this value is close to zero the actual orbit might be an ellipse, a parabola or a hyperbola. In order to find the nature of the orbit, several more observations are required, but even then the orbit often remains undetermined.

Besides these practical aspects, the theoretical implications might be mentioned briefly. We speak about instability when slight changes in the initial conditions result in greatly different orbits. This is the case of orbits for which the constant of energy is close to zero since a small increase in velocity will result in an escaping (hyperbolic) orbit, and a small speed reduction gives an elliptic orbit. Such a case might lead us to the problem of nonpredictability of orbits. The situation is significantly complicated if some other forces enter the system besides those considered in the problem of two bodies, such as drag (which will slow down the orbit), propulsion (which might change the energy to positive or to negative values), perturbations due to other bodies, etc.

For these theoretical reasons, parabolic orbits are of special interest because they represent examples of non-predictable orbits. Their practical significance is limited since they exist only for a highly special value of the energy.

These remarks concerning the examples of the limits of predictability should be compared to the notes appearing in the section on physical constants in the Appendix.

The basic ideas are discussed by Gauss (1809) and Herget (1948). Many examples and additional geometrical presentations are given in McCuskey's (1963) and Thomson's (1961) books. The special case of straight-line motion ($e = 1, b = 0$) is discussed systematically and in detail by Roy (1978).

EXAMPLES

1. The following example shows how the orbital elements of the hyperbolic orbit are established and how time computations are performed.

    Consider the perigee velocity $v_P = 12 \, km/sec$ at an altitude of $h = 1000 \, km$. First the orbit must be classified by using the energy equation (Eqn. 3.19):

    $$v^2 = \frac{2\mu}{r} + k .$$

    After substituting $v = v_P$ and $r = r_P = h + R_E$ we obtain for the constant of energy

    $$k = v_P{}^2 - \frac{2\mu}{h + R_E} = 35.95 \, km^2/sec^2 .$$

    Note that the total energy of the probe per unit mass is $k/2$ since the energy is

    $$H = \frac{1}{2} mv_P{}^2 - \frac{m\mu}{h + R_E} .$$

    Since energy is positive the orbit is a hyperbola. The perigee distance is

    $$h + R_E = a(e - 1)$$

    and the semi-major axis is

    $$a = \mu/k .$$

    From these equations we have $a = 11{,}087.62 \, km$ and $e = 1.66544$. The semi-major axis and the semi-latus rectum are

    $$b = a\sqrt{e^2 - 1} \quad \text{and} \quad p = a(e^2 - 1) ,$$

    or $b = 14{,}766.48 \, km$ and $p = 19{,}665.98 \, km$.

    The angle of the asymptotes is $\alpha = \arctan\sqrt{e^2 - 1} = 53°5'5''$.

Note that by means of the energy equation the velocity may be computed at any point of the hyperbola:

$$v = \sqrt{\mu} \left[\frac{2}{r} + \frac{1}{a} \right]^{1/2} .$$

For instance the velocity at infinity is obtained by the limit process, $r \rightarrow \infty$. We have

$$v_\infty = \left[\frac{\mu}{a} \right]^{1/2} = 5.9958 \; km/sec .$$

The circular velocity at the given altitude is

$$v_c = \left[\frac{\mu}{r_p} \right]^{1/2} = 7.3501 \; km/sec ,$$

which is of course smaller than the perigee velocity.

The escape velocity at the altitude of $h = 1000 \; km$ is

$$v_e = \left[\frac{2\mu}{r_p} \right]^{1/2} = 10.3947 \; km/sec ,$$

which is smaller than the given perigee velocity. This is another way to establish the fact that the orbit is a hyperbola.

The velocity at the point where the semi-latus rectum intersects the hyperbola ($r = p$ and $f = \pi/2$) is

$$v_Q = \left[\mu(\frac{2}{p} + \frac{1}{a}) \right]^{1/2} = 8.7457 \; km/sec .$$

The computed velocities in this case might be ordered as

$$v_P > v_e > v_Q > v_c > v_\infty ,$$

but note that depending on the value of the eccentricity these inequalities might change their order.

To evaluate the time of travel on a hyperbolic orbit, Eqn. (7.5) is used. For instance from perigee (P) to point Q we have

$$t = \frac{1}{n} \left[e \, (\sinh F) - F \right] ,$$

where n is obtained from the relation $n = \sqrt{\mu/a^3}$, and the values of the hyperbolic eccentric anomaly F is computed from Eqn. (7.4). At point Q the true anomaly is 90° or $f = \pi/2$ and F is to be computed from

$$\tan(\pi/4) = \left[\frac{e+1}{e-1}\right]^{1/2} \tan(F/2) \, ,$$

or

$$F = \ln\left[\frac{\sqrt{e+1}+\sqrt{e-1}}{\sqrt{e+1}-\sqrt{e-1}}\right] = \ln(e + \sqrt{e^2-1}) = 1.0976918 \, .$$

The value of F may also be obtained from using Eqn. (7.3). The left side of this equation for $f = \pi/2$ gives $a(e^2-1)$, which is the length of the semi-latus rectum. From Eqn. (7.3) we have

$$e^2 - 1 = e \cosh F - 1 \text{ or } \cosh F = e$$

and consequently $F = \ln(e + \sqrt{e^2-1})$, which is the same result as obtained above.

The "hyperbolic mean motion" becomes $n = 1.94677 \; rad/hour$, and Eqn. (7.5) gives $t = 0.575487 \; hour = 34 \; min \; 31.75 \; sec$.

Note that when Eqn. (7.5) is used, the term $\sinh F$ becomes in our case $\sqrt{e^2-1}$ since $\cosh^2 F - \sinh^2 F = 1$ and $\cosh F = e$.

A quick approximate computation of the travel time uses the average velocity as the perigee and at point Q:

$$v_{ave} = \frac{1}{2}(v_Q + v_P) = 10.37 \; km/sec \, .$$

The distance between points P and Q is

$$\overline{PQ} = \left[\overline{FP}^2 + \overline{FQ}^2\right]^{\frac{1}{2}} \, ,$$

which after substitutions for $\overline{FP} = a(e-1)$ and for $\overline{FQ} = a(e^2-1)$ becomes $\overline{PQ} = r_P\sqrt{1+(e+1)^2} = 21,004.5 \; km$. The time to travel this distance with the above average velocity is $t' = \overline{PQ}/v_{ave} = 0.5625 \; hours$. Note that the error $(t - t')$ amounts to 2.3 percent which supports the idea of approximations, especially considering the fact that no hyperbolic functions were used to obtain the approximate result.

Another question we might ask in connection with our example is related to the design of lunar trajectories. Our probe moving in a hyperbolic orbit will be influenced mostly by the Earth's gravitational field unless it approaches another celestial body such as the Moon. If the departure time is properly selected the probe will approach the Moon, and the Moon's gravitational field will control part of the trajectory. In orbit mechanics we refer to this formulation as the restricted problem of three bodies: the Earth, the Moon, and the probe.

The "restriction" comes from the fact that the probe, because of its small mass, does not influence the motion of the Earth and of the Moon, but the Earth and the Moon control the motion of the third body which is the probe. In Chapter 12 this problem will be discussed in detail, but at this time we will neglect the Moon's gravitational influence on the probe and we will establish the time it takes to reach the distance of the Moon using our hyperbolic orbit. This transfer time will be compared to the time it takes to reach the lunar orbit on an elliptic Hohmann transfer orbit.

For the Earth-Moon distance we will use the semi-major axis of the lunar orbit around the Earth $R_{EM} = 384,000 \, km$ which is only an approximation since the eccentricity of the lunar orbit is 0.0549.

Computation of the hyperbolic transfer time requires the values of a, e, and R_{EM}. Using the same hyperbolic orbit as before, we have to compute the value of F from $R_{EM} = a(e \cosh F - 1)$ and then the transfer time from the hyperbolic Kepler equation. In this way we obtain

$$\cosh F = \frac{R_{EM} + a}{ae} = 21.41735$$

and $F = 3.7568$. Using the same value as before for n we obtain for the transfer time 16.3725 hours. Note that actual transfer time will be shorter since the Moon's gravitational effect will assist the trajectory. (The above computed time corresponds to the time reaching the lunar distance.)

It is of considerable interest to compare this time to an elliptic transfer time using a Hohmann orbit with a circular parking orbit at altitude $h = 1000 \, km$ as before. The departure velocity will be determined so that the transfer orbit will be an ellipse with perigee altitude $R_E + h$ and apogee height R_{EM}. The semi-major axis of the transfer ellipse is

$$a = \frac{R_{EM} + R_E + h}{2} = 195,889 \, km.$$

and the transfer time becomes

$$T = \pi \left[\frac{a^3}{\mu} \right]^{1/2} = 119.84 \; hours \; ,$$

which is approximately 5 days. (Note that the hyperbolic transfer time was about 0.7 day.)

The arrival velocity at the Moon (once again neglecting the lunar gravitational effect) is

$$v_A = \left[2\mu \frac{R_E + h}{R_{EM}} \right]^{1/2} \frac{1}{\sqrt{R_{EM} + R_E + h}} = 0.1976 \ km/sec \ ,$$

and the elliptic perigee velocity is

$$v_P = \left[2\mu \frac{R_{EM}}{R_E + h} \right]^{1/2} \frac{1}{\sqrt{R_{EM} + R_E + h}} = 10.296 \ km/sec \ .$$

The hyperbolic departure velocity is 12 km/sec and the arrival velocity is $v_A = 6.1664 \ km/sec$ which is obtained from the equation of energy, using $r = R_{EM}$. The circular velocity at $h = 1000 \ km$ is 7.3501 km/sec, and the required velocity boost for the elliptic orbit at the perigee is $v_P - v_c = 2.946 \ km/sec$. The circular velocity of the Moon is

$$v_{cM} = \left[\frac{\mu}{R_{EM}} \right]^{1/2} = 1.018 \ km/sec \ ;$$

therefore, the vehicle arriving on an elliptic orbit would require a velocity boost of $v_{cM} - v_A = 0.8207 \ km/sec$ to keep up with the Moon. The probe arriving on a hyperbolic orbit has a higher velocity than the lunar velocity. The velocity vector of the probe arriving on the elliptic transfer orbit is tangential to the Moon's orbit, but this is not the case for the hyperbolic transfer orbit.

Some details of the intersection of the hyperbolic transfer orbit and the Moon's orbit are shown on Figure 7.2. The two components of the arrival velocity of the probe are the radial $\dot{r}$ and the normal $r\dot{f}$ components. The normal component of the arrival velocity may be obtained from the momentum conservation, $c = r^2\dot{f}$ as $r\dot{f} = c/r$ where $c = v_P r_P$ and $r = R_{EM}$. In this way we have $r\dot{f} = 0.23033 \ km/sec$. The direction of the arrival velocity (v_a) is given by the angle between v_a and $\dot{r}$

$$\lambda = \arcsin r \frac{\dot{f}}{v_a} = 2.1406° \ .$$

The angle between the Moon's velocity v_{cM} (which is normal to r) and the probe's velocity which is tangential to the hyperbola is $90° - \lambda$.

The hyperbolic orbit will intersect the lunar orbit when $F = 3.7568$ at point M as shown before. From this the true anomaly becomes $f = 124.731°$ which may be obtained from Eqn. (7.4) or

Eqn. (7.3) by computing f from

$$R_{EM} = \frac{a(e^2 - 1)}{1 + e \cos f} \, .$$

The angle between the asymptote and the radial direction is $180 - (\alpha + f) = 2.1706°$.

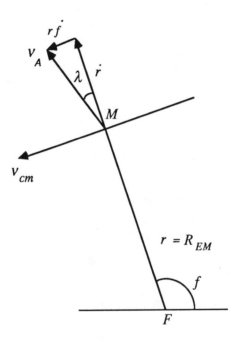

Figure 7.2. Arrival at the Moon on a Hyperbolic Trajectory.

2. The perigee velocities of elliptic and hyperbolic orbits having the same semi-major axes are related by

$$\frac{(v_e)_P}{(v_h)_P} = \left[\frac{e_h - 1}{1 - e_e} \right]^{1/2} \left[\frac{1 + e_e}{1 + e_h} \right]^{1/2} \, ,$$

where $(v_e)_P$ and $(v_h)_P$ are the perigee velocities of elliptic and hyperbolic orbits and e_e and e_h are the corresponding eccentricities. To show the above relation consider the principle of energy conservation which gives for elliptic orbits

$$v_e = \sqrt{\mu}\left[\frac{2}{r} - \frac{1}{a}\right]^{1/2}$$

and for hyperbolic orbits

$$v_h = \sqrt{\mu}\left[\frac{2}{r} + \frac{1}{a}\right]^{1/2}.$$

The perigee distances are

$$r_e = a_e(1 - e_e) \text{ and } r_h = a_h(e_h - 1).$$

The perigee velocities become

$$(v_e)_P = \left[\frac{\mu}{a_e}\right]^{1/2}\left[\frac{1 + e_e}{1 - e_e}\right]^{1/2}$$

and

$$(v_h)_P = \left[\frac{\mu}{a_h}\right]^{1/2}\left[\frac{e_h + 1}{e_h - 1}\right]^{1/2}.$$

If $a_h = a_e$, the required result is obtained from the last two equations. If the eccentricities are related by

$$e_e = 1 - x, \ e_h = 1 + x,$$

where $0 \le x \le 1$, then the inequality $v_e < v_h$ is satisfied. The physical meaning of this is that if the eccentricity of the elliptic orbit is less than 1 by the quantity x and the eccentricity of the hyperbolic orbit is larger than 1 by the same amount, then the hyperbolic perigee velocity is larger then the corresponding elliptic perigee velocity, as expected. For instance, if $x = 0.5$ and $e_e = 0.5$, $e_h = 1.5$ we have for $a_h = a_e$ that

$$\frac{(v_e)_P}{(v_h)_P} = \sqrt{0.6} \text{ and } (v_e)_P < (v_h)_P.$$

The proof of the above result is obtained by substituting $e_e = 1 - x$ and $e_h = 1 + x$ into the equation given for $(v_c)_P/(v_h)_P$. In this way we obtain

$$\frac{(v_e)_P}{(v_h)_P} = \left[\frac{2 - x}{2 + x}\right]^{1/2},$$

which quantity is always less than 1.

Note that if the eccentricities are close to one, i.e. x is much less than 1, the velocity ratio becomes approximately $1 - x/2$.

3. In case the reader is interested in additional numerical examples, the following results are offered without details.

If the circular parking orbit has a period of 2 hours, then the altitude is 1681 km, and the circular velocity is 7.033 km/sec. The escape velocity at this altitude is 9.946 km/sec. To obtain a hyperbolic transfer orbit to the Moon let us use a velocity which is 10 percent higher than the escape velocity, i.e. 10.941 km/sec.

To establish the transfer hyperbola we now use the above altitude of 1681 km and the velocity of 10.941 km/sec which is tangential to the parking orbit at the time of departure. The semi-major axis of the hyperbolic orbit is 19,177.6 km, its eccentricity is $e = 1.42$, its semi-minor axis is 19,340.4 km and the velocity at infinite distance becomes 4.56 km/sec. The transfer time is 20 hours 34 min 29 sec. The angle of the asymptotes is $\alpha = 45.233°$.

4. The reader who likes to manipulate equations will enjoy the following exercise.

For a probe near the Earth we have a tangential velocity v_P (normal to the radius vector from the center of the Earth) at an altitude of h_P. First we establish the ranges of the v_P values which result in elliptic, parabolic and hyperbolic orbits. Using $r_P = h_P + R_E$ for the distance from the center of the Earth to the probe, the energy conservation gives

$$v_P{}^2 - 2\frac{\mu}{r_P} = k .$$

Elliptic orbits correspond to $k < 0$, parabolic orbits to $k = 0$ and hyperbolic orbits to $k > 0$. Consequently for elliptic orbits

$$0 \le V_P < \left[2\frac{\mu}{r_P} \right]^{1/2} ,$$

for parabolic orbits

$$V_P = \left[2\frac{\mu}{r_P} \right]^{1/2} ,$$

and for hyperbolic orbits

$$\left[2\frac{\mu}{r_P}\right]^{1/2} < V_P \; .$$

As a second part of this exercise we establish the semi-major axes, eccentricities and semi-latus rectums (recta, if the reader is a Latinist) as functions of v_P, μ, and r_P for the elliptic, parabolic, and hyperbolic orbits.

For the elliptic orbits

$$a = \frac{\mu r_P}{2\mu - r_P v_P^2} \; ,$$

$$e = \frac{r_P v_P^2}{\mu} - 1 \; ,$$

and

$$p = \frac{r_P^2 v_P^2}{\mu} \; .$$

The first result follows from the energy conservation equation:

$$v_P^2 = \mu(\frac{2}{r_P} - \frac{1}{a}) \; ,$$

the second from using $r_P = a(1 - e)$ and the third from $p = a(1 - e^2)$.

For parabolic orbits $a = \infty, e = 1, p = 2r_P$.

For hyperbolic orbits

$$a = \frac{\mu r_P}{r_P v_P^2 - 2\mu} \; ,$$

$$e = \frac{r_P v_P^2}{\mu} - 1 \; ,$$

and

$$p = \frac{r_P^2 v_P^2}{\mu} \; .$$

As the third part of this exercise we find the velocities for the three cases at point Q (see Figs. 4.1 and 7.1). Note that point Q is located at the intersection of the semi-latus rectum with the orbit.

The velocity at Q is obtained from the energy conservation equation, which for elliptic orbits becomes

$$v_Q{}^2 = \mu \left[\frac{2}{p} - \frac{1}{a} \right] .$$

The result for elliptic orbits is

$$v_Q{}^2 = \frac{2}{r_P} \left[\frac{\mu}{r_P v_P{}^2} - 1 \right] + v_P{}^2 ,$$

which is the same for hyperbolic orbits.

For parabolic orbits we have

$$v_Q{}^2 = \frac{\mu}{r_P} .$$

The last part of this exercise is the computation of the elapsed times between perigee passage and arrival at point Q.

For elliptic orbits Kepler's equation gives

$$nt = E - e \, \sin E ,$$

where the mean motion is given by $n = \sqrt{\mu/a^3}$. The eccentric anomaly (E) at perigee is zero and at point Q is obtained from the value of the true anomaly $(f = \pi/2)$ using Eqn. (5.4).

In this way we have at point Q

$$E = \arcsin \sqrt{1 - e^2}$$

and the transfer time becomes

$$T = \left[\frac{a^3}{\mu} \right]^{1/2} \left[\arcsin \sqrt{1 - e^2} - e \sqrt{1 - e^2} \right] .$$

In this result a and e are to be expressed as functions of μ, r_P and v_P as obtained in the second part of this exercise.

For parabolic orbits the required result is given by Eqn. (7.8) :

$$\left[\frac{\mu}{p^3} \right]^{1/2} t = \tan\frac{f}{2} + \frac{1}{3}\tan^3\frac{f}{2} .$$

Once again, at perigee $f = 0$, $t = 0$, and at point Q, $f = \pi/2$, therefore

$$T = \frac{8}{3} \left[\frac{2r_P{}^3}{\mu} \right]^{1/2} .$$

The case of hyperbolic orbits is similar to the one discussed in connection with elliptic orbits, but now Kepler's equation becomes

$$nt = e\,(\sinh F) - F \ ,$$

where the mean motion is still $\sqrt{\mu/a^3}$ and the hyperbolic eccentric anomaly F is related to the true anomaly by Eqn. (7.3) :

$$\frac{p}{1 + e\,\cos f} = a\,(e\,\cosh F - 1) \ .$$

At the perigee $f = 0$, and from Eqn. (7.3) we have $\cosh F = 1$ or $F = 0$.

At point Q, $f = \pi/2$ and $\cosh F = e$ or $\sinh F = \sqrt{e^2 - 1}$. Using the inverse hyperbolic cosine function, we have

$$T = \left[\frac{a^3}{\mu} \right]^{1/2} \left[e\,\sqrt{e^2 - 1} - \ln(e + \sqrt{e^2 - 1}) \right] \ .$$

In this result a and e are still to be expressed as functions of r_p, v_p and μ. These relations were found in the second part of this exercise.

PROBLEMS

1. Derive formulas for the radial ($\dot{r}$) and normal ($r\dot{\phi}$) velocity components as functions of μ, the semi-major axis, eccentricity and true anomaly for elliptic and hyperbolic orbits.

2. Derive the formula for the velocity $v = \sqrt{\dot{r}^2 + (r\dot{\phi})^2}$, also as a function of μ, a, e and ϕ, and compare the result obtained from the equation of energy, once again for elliptic and hyperbolic orbits.

3. Derive formulas for dr/df, dr/dE, and dr/dl for elliptic and hyperbolic motions and compute their values at the pericenter.

4. A space probe is launched vertically up, from the surface of the Earth, with speed u. Neglecting drag, compute its height, when its velocity becomes zero, as a function of u. When the probe reaches this height, it is given a transverse velocity, v. Find the nature and the parameters of the orbit and show how these depend on u and v.

Chapter 8. Lambert's Theorem

In 1761 Lambert gave his theorem for the initial orbit determination of comets and offered a geometrical proof. This was followed by Lagrange's analytic proof in 1778. The reason to include this more than 200-year-old theorem is its importance to present and future orbit determination and guidance problems concerning moving targets and its use for rendezvous operations. Indeed, the Voyager interplanetary mission, the guidance of the Apollo project and many other space missions utilized the "Lambert Guidance Program." This important subject will be presented only as far as the basic principles are concerned and the reader will be referred to the literature at the end of this chapter.

The theorem states that the duration of time, $t_2 - t_1$, moving on an elliptic orbit from point P_1 to P_2 can be expressed as a function of the semi-major axis of the ellipse, of the sum of the two radial distances between the focus and P_1 and P_2, respectively, and of the length of the chord connecting P_1 and P_2.

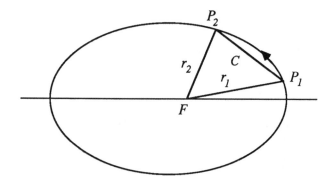

Figure 8.1. Lambert's Theorem.

Analytically the theorem may be written as

$$t_2 - t_1 = f(a, r_1 + r_2, C) ,$$

when the notation of Figure 8.1 is used.

The proof of the theorem is based on the introduction of the angles α and β which are defined by

$$\sin\frac{\alpha}{2} = \frac{1}{2}\left[\frac{r_1 + r_2 + C}{a}\right]^{1/2}$$

and

$$\sin\frac{\beta}{2} = \frac{1}{2}\left[\frac{r_1 + r_2 - C}{a}\right]^{1/2} .$$

The radial distances are given by

$$r_1 = a(1 - e \cos E_1) ,$$

$$r_2 = a(1 - e \cos E_2) ,$$

and from Kepler's equation we have

$$n(t_2 - t_1) = E_2 - E_1 - e(\sin E_2 - \sin E_1) .$$

The chord may be expressed by

$$C^2 = a^2(\cos E_2 - \cos E_1)^2 + (1 - e^2)(\sin E_2 - \sin E_1)^2 .$$

Substituting in Kepler's equation gives Lambert's theorem:

$$n(t_2 - t_1) = \alpha - \beta - (\sin\alpha - \sin\beta) , \qquad (8.1)$$

where $n = \sqrt{\mu/a^3}$.

The elapsed time, $(t_2 - t_1)$, therefore is expressed as a function of $r_1 + r_2, C$ and a.

Note that the equation representing Lambert's theorem does not have a unique solution and its proper use requires careful attention.

Important practical applications are numerous. We start with a simple example and relate the elapsed time between perigee ($f = 0$) and $f = \pi/2$, that is where the semi-latus rectum intersects the ellipse. These are points $P = P_1$ and $Q = P_2$ as shown on Figure 4.1.

The radial distances are

$$r_1 = r_p = a(1 - e)$$

$$r_2 = p = a(1 - e^2),$$

and the chord is

$$C = \sqrt{r_1^2 + r_2^2}$$

or

$$C = a(1 - e)\sqrt{2 + 2e + e^2}.$$

Note that the values of the true anomaly are $f_1 = 0$ and $f_2 = \pi/2$ and the corresponding values of the eccentric anomalies are $E_1 = 0$ and $E_2 = \arccos e$. This latter value may be obtained either from

$$r_2 = a(1 - e\, \cos E_2) = p,$$

or from

$$\tan \frac{f_2}{2} = \left[\frac{1 + e}{1 - e} \right]^{1/2} \tan \frac{E_2}{2}.$$

Using Kepler's equation the elapsed time becomes

$$t_2 - t_1 = \frac{1}{n} \left(\arccos e - e\sqrt{1 - e^2} \right).$$

The same result may be obtained by using Lambert's formula, after evaluating α and β and substituting in Eqn. (8.1).

A similar formulation may be derived for hyperbolic orbits. The starting parameters are the sum of the radial distances, the length of the chord, and the semi-major axis of the hyperbola. The radial distances from Eqn. (7.3) are

$$r_1 = a(e\, \cosh F_1 - 1)$$

and

$$r_2 = a(e\, \cosh F_2 - 1).$$

The terms $\sin(\alpha/2)$ and $\sin(\beta/2)$ used for the elliptic problem now become the corresponding hyperbolic functions:

$$\sinh \frac{\gamma}{2} = \frac{1}{2} \left[\frac{r_1 + r_2 + C}{a} \right]^{1/2}$$

and

$$\sinh\frac{\delta}{2} = \frac{1}{2}\left[\frac{r_1 + r_2 - C}{a}\right]^{1/2}.$$

Kepler's equation for hyperbolic orbits (Eqn. 7.5) becomes

$$n(t_2 - t_1) = e(\sinh F_2 - \sinh F_1) - (F_2 - F_1),$$

and the length of the chord may now be written as

$$C = a\left[(\cosh E_2 - \cosh E_1)^2 + (e^2 - 1)(\sinh E_2 - \sinh E_1)^2\right]^{1/2}.$$

Lambert's theorem for hyperbolic orbits becomes

$$n(t_2 - t_1) = \sinh\gamma - \sinh\delta - (\gamma - \delta). \tag{8.2}$$

Lambert's theorem for parabolic orbits was given by Newton and by Euler. It is even simpler than the above two cases. The time difference becomes

$$t_2 - t_1 = \frac{1}{6\sqrt{\mu}}\left[(r_1 + r_2 + +C)^{3/2} - (r_1 + r_2 + -C)^{3/2}\right]. \tag{8.3}$$

The sign of the second term becomes positive if the angle between r_1 and r_2 is larger than 180°, and it is negative, as given in Eqn. (8.3), when $f_2 - f_1 < 180°$.

In conclusion of this chapter we note that Lambert's theorem can also be used to determine the radius of curvature of the orbit and in this way to find whether a planet is superior or inferior, i.e. whether the Earth or the planet is nearer to the Sun.

One of the outstanding, detailed treatments of Lambert's theorem is given in Plummer's book (1918). For applications to orbit determination techniques, see Bate, Mueller and White (1971), where Lambert's problem is referred to as Gauss' problem. (Note that the year Gauss was born, Lambert died - 1777.) Several practically useful modifications of the original formulation are offered in this reference, such as rendezvous, intercept, etc. Special attention is directed to the astrodynamics applications of the f and g series (see Chapter 3, Example 3), in connection with the "Lambert-Gauss" problem. Important and recent applications can be found in Battin's 1964 book (Chapters 3 and 5) and in his 1987 book (Chapter 6), mentioned in the Appendix.

EXAMPLES

1. As a simple exercise, the previously computed time of travel on a
    hyperbolic orbit shown on Figure 7.1 between points P and Q may be
    computed using Lambert's formulation. The radial distances in case
    of a hyperbolic orbit with perigee altitude $h = 1000$ *km* and velocity
    $v_p = 12$ *km/sec* are

    $$r_1 = r_p = h + R_E = 7378.14 \ km \, ,$$

    and

    $$r_2 = p = a(e^2 - 1) = 19{,}666 \ km.$$

    The previously obtained values for the semi-major axis and
    eccentricity were $a = 11{,}087.62 \ km$, and $e = 1.66544$. The length of
    the chord is

    $$C = \sqrt{r_1^2 + r_2^2} = 21{,}004.5 \ km \, ,$$

    and $\sinh(\gamma/2) = 1.0409$, $\sinh(\delta/2) = 0.3690$. The corresponding
    values for γ and δ may be obtained using the previously given
    logarithmic formulas: $\gamma = 1.8199$ and $\delta = 0.7223$. Substitution in
    Eqn. (8.2), using $n = 1.947$ *rad/hour*, as before, gives
    $t_2 - t_1 = 0.575$ *hour*.

PROBLEMS

1. Solve the Exercises of Chapter 7 using Lambert's theorem.

2. If the area swept out by the radius vector is A_1 and the corresponding
    triangular area enclosed by r_1, r_2 and C is A_2, find A_1/A_2 (see Fig.
    8.1).

Chapter 9. Orbital Elements

The problem of two bodies is represented by a second order differential equation expressing the behavior of the relative position vector of one body with respect to the other (Eqn. 3.8). This vector equation may be written as a set of three second order scalar differential equations which is equivalent to a sixth order system (Eqns. 3.9 - 3.11). The solution (or the orbit) is determined if six initial conditions are given, usually the three components of the relative velocity vector. An alternate set of six parameters which determine the orbit is the subject of this section.

We have seen that the plane in which the motion takes place is fixed since the vector-product of the relative position vector and the relative velocity vector is constant (Eqns. 3.21 - 3.23). Consequently, the motion is always two-dimensional. The geometry or the shape of the orbit is described by two parameters, the semi-major axis and the eccentricity. The orientation of the semi-major axis in the orbital plane is given by ω which is known as the argument of the periaxis and which is the angle between the line of nodes and the semi-major axis as shown on Figures 4.1 and 9.1. The orientation of the orbital plane is determined by two independent parameters, which are the two components of the unit vector normal to the orbital plane,

$$\bar{g} = \frac{\bar{c}}{|\bar{c}|} = \frac{\bar{r}_{12} \times \dot{\bar{r}}_{12}}{|\bar{r}_{12} \times \dot{\bar{r}}_{12}|} .$$ (9.1)

Note that $g_1^2 + g_2^2 + g_3^2 = 1$, therefore only two components are needed. The orbital plane might also be defined by two angles. These angles are the angle of inclination i and the longitude of the ascending node Ω. The relations between these angles and the components of the above-given vector are

$$g_1 = \sin\Omega \, \sin i \, , g_2 = -\cos\Omega \, \sin i \, , g_3 = \cos i .$$ (9.2)

Figure 9.1 shows the orbital elements referred to the equatorial plane which contains the line from the center of the Earth pointing to the Vernal Equinox. This line is the intersection of the equatorial plane and the plane of the ecliptic (the plane of the orbit of the Earth around the Sun). From this direction (γ) the angle Ω is measured counterclockwise. The orbit intersects the equatorial plane at two points, at the ascending node and at the descending node.

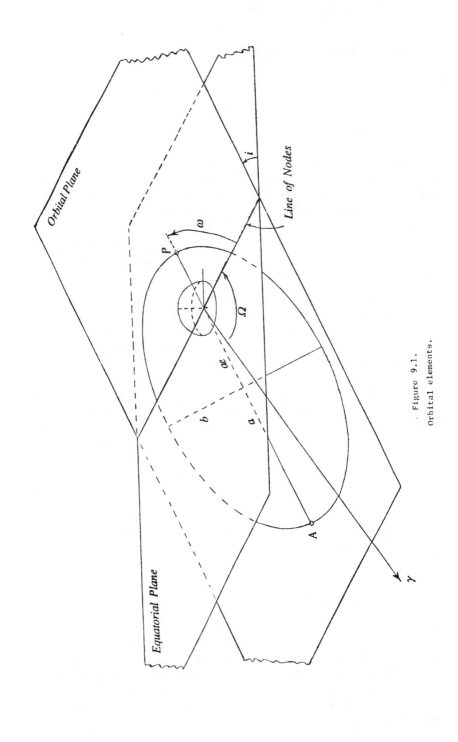

Figure 9.1.
Orbital elements.

The line connecting these points contains the center of the Earth, and it is called the line of nodes. The line pointing to the Vernal Equinox and the line of nodes are both in the equatorial plane, and they determine the angle Ω. The semi-major axis and the line of nodes are both in the orbital plane, and they determine the angle ω known as the argument of perigee, mentioned before. The sum of these two angles is denoted by $\bar{\omega} = \Omega + \omega$ which is called the longitude of perigee. Its significance is that for equatorial orbits, when $i = 0$, the angles Ω and ω are not determined, since there is no nodal line and the orientation of the semi-major axis is given by $\bar{\omega}$. Note that the original notation for the longitude of perigee, "curly π," is substituted here by omega bar.

The orbital elements are a, e, i, ω, Ω and T. The first two determine the geometry or shape of the orbit, and the next three give the orientation. The time of perigee passage, T, determines the location of the body on its elliptical orbit at any given time. The actual determination of the orbital elements are shown in the examples at the end of this chapter and in Chapter 11.

It is natural to inquire about the usefulness of orbital elements as compared to using a Cartesian rectangular coordinate system. The fact that the orbital elements do not vary (in the case of the problem of two bodies) is basically the answer to the above inquiry. Only the time variation is expressed by an angle (such as the true anomaly), as opposed to the variation of the three coordinates when a rectangular coordinate system is used to describe the motion.

The major advantage of using orbital elements is, surprisingly, in problems where these elements are not constant. If the motion of two bodies encounters perturbations, i.e., if other forces are acting in addition to the two-body gravitational interactions, the use of orbital elements becomes mandatory for comprehending the behavior of the system.

The powerful and advanced techniques associated with perturbations are discussed in Chapter 10.

This chapter is concluded with a short remark of some practical interest showing a simple method to compute the components of the velocity vector for two-body motion.

The radial ($\dot{r}$) and normal ($r\dot{\phi}$) velocity components for the two dimensional two-body problem are

$$\dot{r} = \left[\frac{\mu}{p}\right]^{1/2} e \, \sin\phi \tag{9.3}$$

and

$$r\dot{\phi} = \left[\frac{\mu}{p}\right]^{1/2} (1 + e \cos\phi) . \tag{9.4}$$

The radial velocity component may be computed from the equation

$$\dot{r} = \frac{dr}{d\phi}\frac{d\phi}{dt} = \frac{ep \sin\phi}{(1 + e \cos\phi)^2} \dot{\phi} ,$$

which may be simplified by using the conservation of the angular momentum,

$$c = r^2\dot{\phi} = \sqrt{\mu p} .$$

The velocity component normal to the radial direction becomes

$$r\dot{\phi} = \frac{\sqrt{\mu p}}{r} = \left[\frac{\mu}{p}\right]^{1/2} (1 + e \cos\phi) .$$

The velocity is $\left[\dot{r}^2 + (r\dot{\phi})^2\right]^{1/2}$ or

$$v = \left[\frac{\mu}{p}\right]^{1/2}\left[1 + e^2 + 2e\cos\phi\right]^{1/2} . \tag{9.5}$$

Note that the same result may be obtained from the energy equation. To show this, consider the equation

$$v = \sqrt{\mu}\left[\frac{2}{r} \mp \frac{1}{a}\right]^{1/2} ,$$

which follows from Eqn. (4.12). Here the plus sign is to be used for hyperbolic and the minus sign for elliptic orbits. For parabolic orbits we have $v = \sqrt{2\mu/r}$. Substituting

$$\frac{1}{r} = \frac{1 + e \cos\phi}{p} ,$$

we have

$$v = \left[\frac{\mu}{p}\right]^{1/2}\left[2 + 2e\cos\phi \pm \frac{p}{a}\right]^{1/2} .$$

For elliptic orbits $p/a = 1 - e^2$ and for hyperbolic orbits $p/a = e^2 - 1$. Making the substitutions, Eqn. (9.5) is obtained.

Note the advantage of using the parameter p. In this way Eqns. (9.3) - (9.5) are valid for elliptic and for hyperbolic motions using $p = a(1-e^2)$ or $p = a(e^2-1)$ respectively. Furthermore, it might be shown that these results also apply for parabolic motion, if the substitutions of $e = 1$, $p = 2r_p$ are made. The velocity for parabolic orbits is

$$v = \left[\frac{2\mu}{p}\right]^{1/2} (1 + \cos\phi)^{1/2} ,$$

which is obtained from Eqn. (9.5) by using $e = 1$, or from the energy equation with $r = p(1 + \cos\phi)^{-1}$ and $a^{-1} = 0$.

The field of orbit determination from observational results is one of the theoretically and practically important areas of celestial mechanics with its scientific origin going back to Gauss or even further. Since there are many excellent textbooks on this subject, the reader will find in the following only examples and two sets of transformation formulas. The first set relates the longitude (λ) and latitude (β) of a point in the ecliptic system to the right ascension (α) and declination (δ) in the equatorial system, using the center of the Earth as the origin of the coordinates and ε as the obliquity of the ecliptic:

$$\cos\delta \, \cos\alpha = \cos\beta \, \cos\lambda$$
$$\cos\delta \, \sin\alpha = \cos\beta \, \sin\lambda \, \cos\varepsilon - \sin\beta \, \sin\varepsilon$$
$$\sin\delta = \cos\beta \, \sin\lambda \, \sin\varepsilon + \sin\beta \, \cos\varepsilon .$$

The second set relates the rectangular coordinates and the corresponding velocity components to the orbital parameters shown on Fig. 9.1. The origin of the rectangular coordinate systems is at the center of the Earth. The x coordinate points to the vernal equinox, the z coordinate is normal to the equatorial plane and the y axis completes a right-handed system. The equations of the transformation are

$$x = r(\cos\Omega \, \cos\phi - \sin\Omega \, \sin\phi \, \cos i)$$
$$y = r(\sin\Omega \, \cos\phi + \cos\Omega \, \sin\phi \, \cos i)$$
$$z = r\sin\phi \, \sin i$$
$$v_x = -\frac{\mu}{c}(A \, \sin\Omega + B \sin \Omega \, \cos i)$$
$$v_y = -\frac{\mu}{c}(A \, \sin\Omega - B \, \cos\Omega \, \cos i)$$
$$v_z = \frac{\mu}{c}B \, \sin i ,$$

where $A = \sin\phi + e \, \sin\omega$
and $B = \cos\phi + e \, \cos\omega .$

The principle references which specialize on orbit determination are the books by Dubyago, Escobal, Gauss, Herget, Battin and Taff, listed in the Appendix. For the statistical approach see B.D. Tapley's articles in *Recent Advances in Dynamical Astronomy* (1973) and *Long-Time Predictions in Dynamics* (1976), D. Reidel Publ. Co., Dordrecht, Holland, and R. Deutsch's *Estimation Theory*, Prentice Hall, Englewood Cliffs, New Jersey (1965).

EXAMPLES

1. The location of an observing station positioned on the surface of the Earth might be defined at any given time by its geocentric position vector $\bar{R}_o$. Consider that the station makes two measurements of a satellite's positions at t_1 and t_2. The corresponding topocentric position vectors are $\bar{r}_1$ and $\bar{r}_2$. The geocentric position vectors of the satellite are $\bar{R}_1 = \bar{R}_{o1} + \bar{r}_1$ and $\bar{R}_2 = \bar{R}_{o2} + \bar{r}_2$. The two vectors $\bar{R}_1$ and $\bar{R}_2$ are in the plane of the orbit and $R_{oi} = R_o(t_i)$. The normal vector $(\bar{n})$ of this plane can be obtained by calculating the cross product of $\bar{R}_1$ and $\bar{R}_2$:

$$\bar{n} = \bar{R}_1 \times \bar{R}_2 .$$

In this way the orientation of the orbital plane is determined and the angles i and Ω can be computed (see Eqns. 9.1 and 9.2).

Assuming an elliptical orbit, the semi-major axis can be computed from Lambert's theorem (see Eqn. 8.1):

$$(t_2 - t_1) \left[\frac{\mu}{a^3} \right]^{1/2} = \alpha - \beta - (\sin\alpha - \sin\beta) .$$

Here α and β are related to the lengths of the vectors $\bar{R}_1$ and $\bar{R}_2$ and to the chord (C) which in our case is

$$C = \left[|\bar{R}_1|^2 + |\bar{R}_2|^2 - 2\bar{R}_1 \cdot \bar{R}_2 \right]^{1/2} .$$

The only unknown quantity in Lambert's equation, therefore, is the semi-major axis.

If the normal vector of the equatorial plane is $\bar{n}_e$, then the direction of the line of nodes is given by

$$\bar{d} = \bar{n}_e \times \bar{n} ,$$

since the line of nodes is the intersection of the orbital and the equatorial planes.

The eccentricity can be obtained from the equation of the elliptic orbit (see Eqn. 4.5):

$$R_{1,2} = \frac{a(1 - e^2)}{1 + e \cos(\phi_{1,2} - \omega)} ,$$

where ω is the argument of perigee. These two equations (for the subscripts 1 and 2) contain two unknowns, e and ω.

Finally the time of perigee passage, T, can be obtained from Kepler's equation:

$$n(t_2 - T) = E_2 - e \ \sin E_2 \ ,$$

where E_2 is obtained from

$$R_2 = a(1 - e \ \cos E_2) \ .$$

In this way all the orbital elements are obtained from the two observations. This rather simplified example of orbit determination will hopefully inspire the reader to look more deeply into this fascinating subject. The various, new, high technology approaches used for detection of satellites and missiles require new analytical and numerical approaches of orbit determination.

2. The various orbital maneuvers discussed in Chapter 6 did not include changes of the orbital plane. In this example we consider a satellite orbit with inclination $i = 60°$, which is to be changed to $i = 0°$, that is, to an orbit in the equatorial plane. The simplest operation is when the impulsive velocity change (Δv) is executed at the time the satellite crosses the equatorial plane, at one of the nodal points. If the velocity of the satellite on its original orbit is $\bar{v}_1$, and after the maneuver is $\bar{v}_2$, our exercise is to find $\Delta \bar{v}$, so that $|\bar{v}_1| = |\bar{v}_2|$.

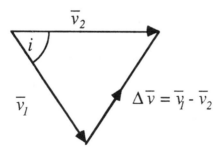

Figure 9.2. Change of Inclination of the Orbital Plane.

As shown on Fig. 9.2,

$$|\Delta \bar{v}| = 2|\bar{v}_1| \ \sin\frac{i}{2} \ .$$

If $i = 60°$, the magnitude of the velocity change is the same as the magnitude of the velocity. This example shows that maneuvers which change the inclination of the orbit of a satellite may require large expenditures in propulsion.

3. In the first example of this chapter the orbital elements were determined from two positional observations, i.e. the six orbital constants of integration were related to the six components of $\bar{r}_1$ and $\bar{r}_2$. Another possibility is to use the six components of the position and velocity vectors, obtained simultaneously, and relate these to the six orbital elements. Both problems can be inverted and either the two position vectors (Example 1) or the state vector (Example 3) can be determined from the orbital elements. If the position and velocity vectors at a given time are $\bar{r}$ and $\dot{\bar{r}}$, the semi-major axis can be obtained immediately from the energy integral:

$$v^2 = \mu \left[\frac{2}{r} - \frac{1}{a} \right] ,$$

where $v = |\dot{\bar{r}}|^2$ and $r = |\bar{r}|$. If the given state vector results in a negative value for a, we conclude that the observed body is moving on a hyperbolic orbit. In what follows we shall assume an elliptic orbit. The angular momentum integral gives

$$\bar{c} = \bar{r} \times \dot{\bar{r}} ,$$

which vector is normal to the orbital plane. Eqns. (9.1) and (9.2) give the elements i and Ω. From the length of $\bar{c}$, the eccentricity can be obtained (Eqn. 4.6):

$$p = c^2/\mu = a(1 - e^2) .$$

An alternate determination of the eccentricity is recommended, especially, when the orbit is near circular. Using the polar equation of an ellipse, we have

$$r = \frac{p}{1 + e \cos f} ,$$

from which

$$e \cos f = \frac{p}{r} - 1 ,$$

and   (9.6)

$$e \sin f = \frac{c}{\mu} \dot{r} .$$

The second equation is obtained from the first one by taking the time derivative of both sides and using the angular momentum integral in its polar form:

$$c = r^2 \dot{f} \ .$$

Equations (9.6) allow the computation of the values of e and f directly from the observational results, since $r = |\bar{r}|$, $p = c^2/\mu$, or $p = |\bar{r} \times \dot{\bar{r}}|^2/\mu$, and $\dot{r} = (\bar{r} \cdot \dot{\bar{r}})/|\bar{r}|$, as mentioned before (see Fig. 3.2).

Note that if the eccentric anomaly E is used instead of the true anomaly, Eqns. (9.6) become

$$e \cos E = 1 - \frac{r}{a} \ ,$$

and   (9.7)

$$e \sin E = \frac{\dot{r}r}{na^2} = \frac{(\bar{r} \cdot \dot{\bar{r}})}{\sqrt{\mu a}} \ .$$

The second equation is obtained by taking the time derivative of the first equation and eliminating $\dot{E}$ by using Kepler's equation,

$$\dot{E} = \frac{an}{r} \ .$$

The mean motion n can be eliminated using Kepler's law.

The advantage of using the eccentric anomaly in the above proceedings instead of the true anomaly is that the mean anomaly at epoch can be immediately obtained from Kepler's equation. The argument of perigee, ω is obtained from the argument of latitude, $\phi = \omega + f$ (see Fig. 4.1) by writing

$$r \cos\phi = \bar{r} \cdot \bar{n} = x \cos\Omega + y \sin \Omega \ ,$$

and

$$r \sin\phi = z \sin^{-1}i \ .$$

Here x, y, z are the components of the position vector $\bar{r}$, and $\bar{n}$ is the unit vector along the line of nodes (see Fig. 9.1).

Chapter 10. Perturbation Methods

Introduction.

After the detailed discussion of the problem of two bodies, we extend the horizon of orbit mechanics and inquire into the cases when the dynamical model increases its complexity. In many practically important problems, two-body orbits are only slightly influenced or perturbed by other effects. In such cases, the conic section orbits are used as the basis and the external effects are considered as perturbations, modifying these conic sections. As mentioned in the conclusion of the previous chapter, the orbital elements will show small variations as long as the perturbations are small.

The subject of perturbations is usually considered beyond the level of introductory treatments. Nevertheless, we include the basic ideas following our established principle of offering the fundamentals instead of complex details. The terminology and some definitions will be introduced such as the concept of osculating orbits, the method of variation of parameters, special and general perturbations, periodic and secular effects of perturbations, etc. This is followed by two simple problems offering the details of the analytic solutions of general perturbation problems in planetary theory. In general, by planetary theory we mean the prediction of the motion of planets in the solar system. The orbits of the planets are governed by the Sun and they perturb each other gravitationally.

This chapter is concluded by describing one of the outstanding and unsolved problems of celestial mechanics, stability.

Applications to the motion of Earth satellites are discussed in Chapter 11.

Basic Concepts and Definitions.

The major perturbations usually are due to the presence of a third body, non-spheriodicity of the participating bodies or atmospheric drag.

We speak about "third body perturbation" when the two-body motion is perturbed gravitationally by an additional third body. For instance a satellite around the Earth with large semi-major axis might be influenced by the Moon and/or by the Sun. Such "deep space probes" are on perturbed elliptic orbits, and their behavior can be best described by giving the variations of the orbital elements of the orbit.

It might be useful at this point to mention the difference between periodic and secular effects of the perturbations. By periodic effects we mean variations of the orbital elements in a periodic fashion, as represented by periodic functions of time. If the semi-major axis of a satellite orbit changes periodically, with a small amplitude, the perturbation does not have significant effects. But if the semi-major axis changes secularly, increasing or decreasing monotonically, the satellite will escape or collide with its planet. Consequently, secular effects are more important in the long-term predictions than periodic effects.

The concept of perturbations is one of the most important ideas in celestial mechanics. Besides the above-mentioned meaning of perturbation, i.e., the small influence in addition to two-body effects, the terms of general and special perturbations should be reviewed. These classical terms are connected with analytical and numerical methods in orbit predictions (or propagations). Analytical results of orbit predictions allow the representation of the solution for many sets of initial conditions. These methods, therefore, are known as "general" perturbation techniques since they give "general" solutions. Numerical integration establishes an orbit for a single set of initial conditions, and it is referred to as "special" perturbation method. Note that general perturbation methods, while giving solutions for many sets of initial conditions, do not offer solutions for all possible initial conditions. This problem is connected with the concept of integrability of dynamical systems as well as with the convergence of series solutions and will be discussed in connection with the problem of three bodies in Chapter 12.

The disadvantages of special perturbation are:

1. if we need the location of a body at time t we have to compute it for times before t (between the initial time t_o and t);

2. no general character of the motion is obtained, i.e. usually no insight is gained concerning the basic properties of the motion;

3. long-time predictions are of questionable validity.

The advantages of special perturbation techniques are:

1. predictions are required usually for a limited time interval for problems of practical interest;

2. no complicated analytical work is required;

3. numerical integration techniques (and computer programs) are available with considerable sophistication using various sets of

variables, applicable to a great variety of practically important problems.

Three well-known methods of perturbations are Cowell's, Encke's and Hansen's. The first is recognized today as a "direct" method, when the differential equations of motion are integrated using rectangular coordinates. Formulation is simple since no distinction is made between the basic two-body forces and the perturbations. Encke's method, on the other hand, uses the idea that in celestial mechanics and in space dynamics the orbits usually can be considered as perturbed two-body orbits. Consequently, Encke's method integrates the difference between a reference orbit (a precomputed osculating orbit, see next section) and the perturbed (actual) orbit. When the deviation between the reference and the perturbed orbits becomes large, a new reference orbit is selected and the process is continued. This step is known as rectification. Note that, in principle, the reference orbit can be any precomputed orbit, and Encke's method can be greatly improved if a reference orbit is used which is closer to the actual orbit than a two-body orbit would be. In general, the computations of satellite and planetary orbits are three to ten times faster using Encke's method as compared to Cowell's. Hansen's method uses polar coordinates and combines the variation of coordinates with the variation of the elements, which will be discussed later on in this chapter.

The conclusion is that, because in general the equations of celestial mechanics are non-integrable, neither general nor special perturbation methods can give solutions of arbitrary long-time validity (see Chapter 12). Here we disagree with Lambert who was convinced that all problems of celestial mechanics can be considered solved since by numerical integration the orbits can be computed. The ignorami of today can be easily recognized when they share Lambert's erroneous opinion.

Comparisons of advantages and disadvantages between general and special perturbation methods strongly depend on the problem to be solved, on the accuracy requirements, on the required length of predictions, on the nature of trajectories, on the selection of the variables, etc.

The most famous general perturbation theory deals with the motion of the Moon around the Earth as perturbed by the Sun. This is known as the Lunar Theory and will be discussed in Chapter 12.

Osculating Orbits.

An important concept of perturbation in orbit mechanics is called "osculating orbit." Besides its importance it is also often misleading and, therefore, careful attention will be given to this basically simple idea.

First, it is recalled that the solutions of the two-body problem are conic sections. For most practical applications these are ellipses which satisfy the differential equation

$$\ddot{\bar{r}} + \mu \frac{\bar{r}}{|\bar{r}|^3} = 0 . \tag{10.1}$$

Using the true anomaly as the independent variable an elliptic orbit is represented by

$$r = \frac{p}{1 + e \cos f} \tag{10.2}$$

as it was shown before.

If there are some perturbations, then Eqn. (10.1) must be modified to

$$\ddot{\bar{r}} + \mu \frac{\bar{r}}{|\bar{r}|^3} = \bar{P} , \tag{10.3}$$

where $\bar{P}$ representing the perturbation might depend in general on $\bar{r}, \dot{\bar{r}}$ and t.

It is noted that no closed form, explicit solution of Eqn. (10.1) exists using time as the independent variable. Consequently, such solutions usually do not exist for Eqn. (10.3). Nevertheless, if $\bar{P}$ is small compared to $\mu \bar{r} |\bar{r}|^{-3}$, the perturbed orbit will not differ much, for a short time, from an "osculating" two-body orbit which exists when $\bar{P} = 0$. When $\bar{P} \neq 0$ the unperturbed two-body orbit will not be the solution of Eqn. (10.3), unless the elements of this orbit change in time.

Consider the initial conditions $\bar{r} = \bar{r}_o$, $\dot{\bar{r}} = \dot{\bar{r}}_o$ at $t = t_o$ and obtain the osculating solution from Eqn. (10.1). The orbital elements will not be constants on the perturbed orbit. At any time, using the initial conditions of the perturbed orbit we can use the solution of Eqn. (10.1) to obtain an osculating orbit. This is an ellipse which describes the perturbed solution at that instant.

Therefore, we define the osculating orbit as the two-body solution which is obtained at any instant if the perturbations are neglected. The elements of this orbit are called the osculating elements which of course change with time for the perturbed orbit.

The advantage of using this concept is that we can visualize how the orbital elements vary. If some of the elements change as periodic functions of time, especially with small amplitude, their effect on the motion usually is not significant. If, on the other hand, one of the elements changes, for instance, linearly with time, its effect will be sooner or later noticeable. This is known as secular variation.

A pertinent example is the variation of the semi-major axis a of the Earth's orbit around the Sun. If a shows only periodic variations, the Earth's orbit does not change significantly. If a increases or decreases with time, say linearly, the Earth's orbit will either depart from the Sun or will approach the Sun. As a increases, the Earth's orbit might be significantly influenced by Mars, Jupiter, etc. If the semi-major axes of the planets show secular changes, the order of the planets might change. This is what is known as the hierarchical instability (Roy, 1982) of the planetary system. Our present day knowledge concerning the solar system excludes the existence of secular variations in the semi-major axes of the planets for 10^8 years.

In summary we state that an osculating orbit is the orbit the body would follow if the perturbation would be turned off.

Variation of Parameters.

We shall now combine three ideas and establish one of the useful methods of perturbation theory, known as the variation of parameters. These three ideas are not new to the readers at this point, and therefore, they might enjoy finding that they are able to solve problems which are not usually included in introductory texts. As usual, the method will be demonstrated by two simple examples.

The basic idea of the variation of parameters method is as follows. The position vector $(\overline{r})$ of a body moving in an arbitrary force field is a function of the time (t) and of the six initial conditions $(c_1, c_2 \cdots c_6)$. The idea of osculating orbits tells us that these six initial conditions can be the six orbital elements of the two-body orbit which would represent the solution without perturbations. The perturbed orbit will have variable orbital elements, as discussed before; therefore, our solution of the perturbed problem is

$$\overline{r} = \overline{F}\ (t, c_1, c_2 \cdots c_6)\ , \tag{10.4}$$

where the orbital elements or parameters $(c_1 \cdots c_6)$ now depend on the time.

The variation of parameters method, therefore, first solves the unperturbed equations and then substitutes this solution into the equations representing the perturbed problem, assuming that now the parameters $(c_1 \cdots c_6)$ are variables. This is of course not a new idea to the reader since introductory courses on differential equations discuss the solution of inhomogeneous, linear differential equations by finding the variations of the constants appearing in the solution of the homogeneous equations. The solution of the inhomogeneous equation will be the sum of the solution of the homogeneous equation plus a particular solution of the inhomogeneous equation.

At this point the question of nonlinearity enters the picture, since our equations in celestial mechanics are basically nonlinear due to the nonlinearity of the gravitational field. We are proposing to use the method of variation of parameters to solve nonlinear differential equations. The answer to this dilemma is along two lines. First, as we have seen, the gravitational two-body problem can be represented by a linear differential equation, using the proper variables. The nonlinearity of the perturbed equation will not disappear, in general, by using new variables. If the perturbation depends only on time, the perturbed equation can be transformed into an inhomogeneous linear differential equation. If the perturbation brings in nonlinearity, we still can use the solution of the unperturbed equation and the method of the variation of parameters to find the solution.

In summary, the three basic ideas mentioned above are osculating orbits, variable orbital elements and linear two-body equations.

Two Examples of Perturbations.

The two examples which follow represent applications of the above ideas. The first reviews the method of variation of parameters as applied to a second order linear inhomogeneous differential equation describing a perturbed harmonic oscillator. The second will use the technique to establish an approximate solution for a second order nonlinear differential equation representing the perihelion change of Mercury.

1. Consider the differential equation describing a one-dimensional harmonic oscillator with time-dependent perturbation:

$$\ddot{x} + x = R(t).\tag{10.5}$$

Note that this perturbed harmonic oscillator for us is a perturbed and transformed two-body problem.

Introducing $x_1 = x$ and $x_2 = \dot{x}$, the original second order differential equation (Eqn. 10.1 or 10.4) can be written as a system of two first order equations:

$$\dot{x}_1 = x_2,$$
$$\dot{x}_2 = R(t) - x_1.\tag{10.6}$$

If $R(t) = 0$, the solution is

$$x_1 = c_1 \sin t + c_2 \cos t,$$

$$\tag{10.7}$$

$$x_2 = c_1 \cos t - c_2 \sin t \, ,$$

where c_1 and c_2 are constants of integration, and depend on the initial conditions. In orbit mechanics, these constants are related to the orbital elements.

If $R(t) \neq 0$, the above solutions can be substituted into Eqns. (10.6) assuming that c_1 and c_2 are time-dependent. The first of Eqns. (10.6) results in

$$\dot{c}_1 \sin t + \dot{c}_2 \cos t + c_1 \cos t - c_2 \sin t = c_1 \cos t - c_2 \sin t$$

or

$$\dot{c}_1 \sin t + \dot{c}_2 \cos t = 0 \, .$$

The second of Eqn. (10.6) gives

$$\dot{c}_1 \cos t - \dot{c}_2 \sin t = R(t) \, .$$

From the last two equations the derivatives of the variable parameters c_1 and c_2 can be obtained:

$$\dot{c}_1 = R(t) \cos t \quad \text{and} \quad \dot{c}_2 = -R(t) \sin t \, .$$

For a given perturbation function, $c_1(t)$ and $c_2(t)$ can be obtained by integration. The above review might assist in following the next practical example.

2. The following example shows, in details, some of the approximations made when the effects of perturbations due to relativity are evaluated in the motion of Mercury's perihelion. This example was selected because of the simple nature of the derivation, allowing us to see the essential aspects without too many complications. The basic theory of general perturbation method will be discussed following this example since in this way the mathematics of this beautiful, classical and somewhat complicated technique will be less confusing to the reader. In Chapter 4, Eqn. (4.2) is a linear, second order differential equation, representing the variation of $u = 1/r$ with the true anomaly, ϕ. The differential equation is

$$\frac{d^2u}{d\phi^2} + u = \frac{\mu}{c^2} \, , \tag{10.8}$$

where $\mu = G(m_1 + m_2)$ and $c = r^2\dfrac{d\phi}{dt}$ is the constant of the angular momentum. The solution of this equation is given as

$$u = \frac{\mu}{c^2} + A \cos(\phi - \phi_o) \ , \qquad (10.9)$$

where A and ϕ_o are the constants of integration, which depend on the initial conditions. Introducing the corresponding orbital elements, we have

$$u = \frac{\mu}{c^2} [1 + e \cos(\phi - \omega)] \ , \qquad (10.10)$$

where e is the eccentricity and ω is the argument of the perihelion.

One of the basic ideas of general perturbation technique is the following. If there are other than the conventional two-body gravitational forces acting, then Eqn. (10.8) is not valid anymore, and neither is the solution, Eqn. (10.10). The question is if a modified form of Eqn. (10.10) can be used as a solution of the modified (perturbed) Eqn. (10.8). Note that in Eqn. (10.10) the orbital elements e and ω are constant, and it seems to be a natural modification of Eqn. (10.10) to consider these elements variable and investigate if in this way the perturbed version of Eqn. (10.8) can be satisfied. Note that we are speaking, once again, about osculating elements since Eqn. (10.10) represents an elliptic orbit at any instant. The elements will be different at another time and the ellipse will change. Note that we involve presently only two elements e and ω, since the solution of our second order differential equation has only two constants of integration.

Let us now follow in detail the above idea, and add the perturbing term P to the right side of Eqn. (10.8). Consider the solution in the form of Eqn. (10.10) with variable e and ω and substitute this into the perturbed equation. In this way, a second order differential equation is obtained for the elements e and ω. The algebra is considerably simplified if our second order differential equation (Eqn. 10.8) is written as two first order differential equations. This is an often used technique of solving higher order differential equations, and it is especially popular in celestial mechanics. Before turning to the algebra outlined above, a few remarks will be added to support the representation of our problem in the form of first order differential equations.

The general three-dimensional formulation of the problem of two bodies, using vectorial notation is

$$\ddot{\bar{r}} = - \mu \frac{\bar{r}}{|\bar{r}|^3} \ , \qquad (10.11)$$

as given in Chapter 3 by Eqn. (3.8), where $\bar{r}_{12}$ is used for $\bar{r}$. This is a

second order differential equation using the vector $\bar{r}$ to be determined
as the function of time. In scalar form Eqn. (10.11) is equivalent to
three second order differential equations as discussed in Chapter 3.
Each of these three equations have two constants of integrations, and
therefore, the whole system of equations has six constants. This is not
surprising since the initial conditions of Eqn. (10.11) are

$$\bar{r}(t_o) = \bar{r}_o \quad \text{and} \quad \dot{\bar{r}}(t_o) = \dot{\bar{r}}_o \ ,$$

or the three position coordinates and the three velocity components.
Another (vectorial) way to look at the situation is to consider Eqn.
(10.11) as a second order differential equation for the position vector
$\bar{r}$. The initial conditions are the two vectors $\bar{r}_o$ and $\dot{\bar{r}}_o$. No matter
how we count, the number of initial conditions and the order of the
system describing the problem is the same.

A three-dimensional orbit is the solution of a second order
differential equation for the vector $\bar{r}$. This system can be written as
six first order differential equations for the scalar components of $\bar{r}$ and
$\dot{\bar{r}}$. For instance, Eqn. (10.11) can be written as

$$\dot{\bar{r}} = \bar{\rho} \ ,$$

and

$$\dot{\bar{\rho}} = - \mu \frac{\bar{r}}{|\bar{r}|^3} \ .$$

These vector equations represent six scalar equations if the
components of $\bar{r}, \bar{\rho}, \dot{\bar{r}}$ and $\dot{\bar{\rho}}$ are used. If these components are
denoted by (x, y, z) for $\bar{r}$; (ξ, η, ζ) for $\bar{\rho}$; $(\dot{x}, \dot{y}, \dot{z})$ for $\dot{\bar{r}}$; and
$(\dot{\xi}, \dot{\eta}, \dot{\zeta})$ for $\dot{\bar{\rho}}$, then the above two first order vector differential
equations might be written as six first order scalar equations:

$$\dot{x} = \xi, \dot{y} = \eta, \dot{z} = \zeta \ ,$$

$$\dot{\xi} = - \mu \frac{x}{r^3}, \dot{\eta} = - \mu \frac{y}{r^3}, \dot{\zeta} = - \mu \frac{z}{r^3},$$

where $r^3 = (x^2 + y^2 + z^2)^{3/2}$.

The six orbital elements discussed before might be looked upon
as another way to describe the six initial conditions.

Returning now to our original problem, we consider the second
order scalar differential equation, Eqn. (10.8), and write it as two first
order differential equations

$$\frac{du}{d\phi} = v$$

and   (10.12)

$$\frac{dv}{d\phi} = -u + \frac{\mu}{c^2} .$$

In case perturbation (P) appears in Eqn. (10.8) we have

$$\frac{d^2u}{d\phi^2} + u = \frac{\mu}{c^2} + P ,$$ (10.13)

or

$$\frac{du}{d\phi} = v$$

and   (10.14)

$$\frac{dv}{d\phi} = -u + \frac{\mu}{c^2} + P .$$

The problem is to solve Eqn. (10.13) or the equivalent system of Eqns. (10.14) as outlined before.

The solution of system (10.12) is

$$u = \frac{\mu}{c^2} [1 + e \cos(\phi - \omega)]$$

and   (10.15)

$$v = -e \frac{\mu}{c^2} \sin(\phi - \omega) ,$$

which last equation is obtained from the first part of Eqn. (10.12).

The idea is now to substitute Eqns. (10.15) into Eqns. (10.14), assuming that e and ω are not constants and to find the functions $e(\phi)$ and $\omega(\phi)$ which satisfy the perturbed equations. When u, as given by the first of Eqns. (10.15), is substituted in the first of Eqns. (10.14), derivatives of e and ω with respect to ϕ must be included:

$$\frac{du}{d\phi} = \frac{\mu}{c^2} [e' \cos(\phi - \omega) - e(1 - \omega') \sin(\phi - \omega)] ,$$

where

$$e' = \frac{de}{d\phi} \quad \text{and} \quad \omega' = \frac{d\omega}{d\phi} .$$

The first of Eqns. (10.14), therefore, becomes

$$-e \frac{\mu}{c^2} \sin(\phi - \omega)$$

$$= \frac{\mu}{c^2} [e' \cos(\phi - \omega) - e(1 - \omega') \sin(\phi - \omega)] ,$$

or

$$e' \cos f + e \omega' \sin f = 0 , \tag{10.16}$$

where $f = \phi - \omega$ is the true anomaly.

The second step consists of the substitution of Eqns. (10.15) into the second of Eqns. (10.14). Note that the derivative of v becomes

$$\frac{dv}{d\phi} = -\frac{\mu}{c^2} [e' \sin(\phi - \omega) + e(1 - \omega') \cos(\phi - \omega)] .$$

The second of Eqns. (10.14), therefore, becomes

$$-\frac{\mu}{c^2} [e' \sin f + e(1 - \omega') \cos f]$$

$$= -\frac{\mu}{c^2} (1 + e \cos f) + \frac{\mu}{c^2} + P ,$$

or

$$-e' \sin f + e \omega' \cos f = \frac{c}{\mu^2} P . \tag{10.17}$$

Eqns. (10.16) and (10.17) now can be solved for e' and ω' with the following result:

$$e' = -\frac{c^2}{\mu} P \sin f ,$$

$$\tag{10.18}$$

$$\omega' = -\frac{1}{e} \frac{c^2}{\mu} P \cos f .$$

It is now necessary to have some information concerning P in order to proceed. As mentioned at the beginning of the discussion of this example, P represents the correction to the Newtonian formulation by the general relativity effect, or

$$P = \frac{\varepsilon}{r^2} = \varepsilon u^2 ,$$

where $\varepsilon = 3\mu/c_o^2$, and c_o is the velocity of light. This unfortunate notation is the consequence of the previously established notation, according to which c represented the angular momentum (as often is the case in celestial mechanics) and not the velocity of light.

Now we substitute

$$P = \varepsilon \frac{\mu^2}{c^4} (1 + e \cos f)^2$$

in Eqns. (10.18) and obtain

$$e' = - \varepsilon \frac{\mu}{c^2} (1 + e \cos f)^2 \sin f ,$$

(10.19)

$$\omega' = \frac{\varepsilon}{e} \frac{\mu}{c^2} (1 + e \cos f)^2 \cos f .$$

Note that $f = \phi - \omega$ and the prime denotes derivatives with respect to ϕ; therefore, the solution of Eqns. (10.19) represents serious analytical problems. Let us now make the approximations that the elements (e and ω) change slowly; therefore, they might be considered constants on the right side of Eqn. (10.19) at least for one revolution of Mercury. This is a reasonable approximation, since the dimensionless constants of the right sides of Eqns. (10.19) have the numerical values:

$$\frac{\varepsilon\mu}{c^2} = \frac{\varepsilon}{p} = 8 \times 10^{-8} \text{ and } \frac{\varepsilon}{e} \frac{\varepsilon}{c^2} = 4 \times 10^{-7} .$$

The variable terms on the right sides are less than $(1 + e)^2$; therefore, the approximation concerning the slowly changing nature of e and ω seems to be reasonable. Let us now integrate Eqns. (10.19) for one revolution of Mercury, assuming that e and ω are constant on the right side of the equation. For the change of eccentricity we have

$$\Delta e = - \varepsilon \frac{\mu}{c^2} \int_0^{2\pi} \left[1 + e \cos(\phi - \omega) \right]^2 \sin(\phi - \omega) \, d\phi ,$$

where $\phi - \omega = f$ and $d\phi = df$, since ω is considered a constant. The result of the integration is

$$\Delta e = \frac{\varepsilon\mu}{3ec^2} \left[(1 + e \cos f)^3 \right]_0^{2\pi} = 0 .$$

For the argument of the perihelion we have

$$\Delta\omega = \frac{\varepsilon}{e} \frac{\mu}{c^2} \int_0^{2\pi} [1 + e \cos(\phi - \omega)]^2 \cos(\phi - \omega) \, d\phi ,$$

or

$$\Delta\omega = \frac{\varepsilon\mu}{ec^2}\left[\sin f + e\left(f + \frac{\sin 2f}{2}\right) + e^2\sin f\left(1 - \frac{\sin^2 f}{3}\right)\right]_0^{2\pi},$$

or

$$\Delta\omega = 2\pi\frac{\varepsilon\mu}{c^2} = 50.19 \times 10^{-8}\, rad/rev.$$

Mercury's period of revolution is 0.24085 year; therefore, the change of ω in 100 years is

$$\frac{50.19 \times 10^{-8}}{0.24085} \times \frac{180}{\pi} \times 3600 \times 100 = 42.983''.$$

Note that computations show that the perturbations of other planets result in 531 sec of arc per century of change in the perihelion of Mercury. Observational results indicate 574 seconds of arc per century and the difference is accounted for by the above evaluated value which is approximately 43"/century.

Note that no secular change of the eccentricity is indicated by the above derivation.

Leverrier was one of the astronomers who is credited with the prediction of the existence and possible location of Neptune from its perturbations of Uranus. He was also concerned with the motion of Mercury and established the theoretical value of 527 seconds/century for the advance of Mercury's perihelion in 1865, due to perturbations of other planets.

Stability.

One of the problems related to the long-time behavior of orbiting bodies is stability, which is an important, difficult and unsolved problem of celestial mechanics. The stability of the rigid body (rotational) motion of satellites and space stations is of considerable interest and it is treated in detail in the literature (see Fitzpatrick, Chapter 14 and Thomson, Chapters 5-7). The following remarks will be concentrating on the orbital stability of natural and artificial bodies.

The reader's attention is directed to the over fifty definitions of various kinds of stability, often leading to contradictory conclusions. It is strongly recommended that prior to any announcement of the stability of a dynamical system, the definitions used be clarified.

An elliptic two-body orbit is generally considered stable since a small change of the initial conditions will not change much the orbital elements, or the shape and orientation of the orbit. This kind of stability, known as orbital stability, is intuitively clear and can be shown analytically without much difficulty. Consider, on the other hand, the same elliptic orbit and let us change again slightly the initial conditions so that the semi-major axis will change ever so slightly. The orbital stability is still valid, but the change of the length of the semi-major axis will result in a change of the period. After the change of the initial conditions, the small change in the mean motion will displace the body along the orbit. After a sufficiently large number of revolutions the disturbed body might be close to apogee at the time the body on its original orbit will be at perigee. The two orbits will be very close but the distance between the body on the original orbit and the body on the slightly changed orbit will be the length of the major axis. This behavior certainly cannot be considered stable in spite of the fact that the original and the new orbits are close. Poincaré (using the first, geometrical idea) calls our motion stable, while Lyapunov (using the second, kinematic idea) considers the motion unstable. Poincaré's definition is known as orbital stability and Lyapunov's is usually referred to as isochronous (equal time or simultaneous) stability.

An unsolved and rather important problem in celestial mechanics is the stability of the solar system. The readers, after the previous example, will require a definition of stability before they would attempt to find the answer. One of the generally accepted definitions of the stability of planetary systems was offered by Laplace (1773) according to which stability requires that the semi-major axes of the planetary orbits show no secular changes, only small periodic changes so that orbits do not intersect. Another, similar definition connects planetary stability with no collisions and no escapes.

During the existence of the solar system (estimated 5×10^9 years) apparently and probably no major changes have occurred, and numerical integrations indicate stability for the next 10^8 years. Problems with the convergence of analytical series-solutions are presently being clarified, but - as humiliating as this is - we must admit that the problem of stability is still unsolved (see Roy's book, Chapter 8).

Concerning the details of the second problem of this chapter, see U. J. J. Leverrier, "Théorie du mouvement de Mercure," Ann. Observ. Imp. Paris (Mém.), Vol. 5, pp. 1-196, 1859 and N. T. Roveveare *Mercury's Perihelion. From Leverrier to Einstein*, Clarendon Press, Oxford (1982). Regarding details of general perturbation methods, see Brouwer and Clemence (1961). A collection of over 50 definitions of stability is given in V. Szebehely, "Review of Concepts of

Stability," Celestial Mechanics, Vol. 34, pp. 49-64, 1984. The original papers concerning Cowell's and Encke's methods are by P. H. Cowell and A. D. Crommelin in the Mon. Not. Roy. Astron. Soc., Vol. 68, p. 576, 1908 and by J. F. Encke in The Berliner Jahrbuch, 1857. P.A. Hansen's method is described in P. Musen's article in the Astronomical Journal, Vol. 63, p. 426, 1958. For a general discussion see Danby's book, pp. 230-238.

Chapter 11. Orbits of Artificial Satellites

Introduction.

In this section the concepts of potential function and potential energy will be clarified as a first step, since these are essential in the study of the effect of the Earth's gravitational field on its artificial satellites. Certain force fields can be obtained by taking the derivative of a function, known as the potential function. These force fields are called conservative, and the field of gravity is one of the important examples. One of the many advantages of this concept is that it is usually easier to represent a force field using its scalar potential, than the vector function describing the actual force field. Another important application appears in the integral of energy, where the sum of the kinetic and potential energies is constant for conservative fields. The relation between the potential function and the potential energy will be also discussed in the following.

The purpose of this book is to teach orbital mechanics and not to dwell upon general concepts of mathematics and of dynamics. When the gravitational field of the Earth is described, we do this in order to understand the motion of the artificial satellites around the Earth. The motion of these satellites is governed by the gravitational field of the Earth which in turn is usually given by its potential function.

If the force field is simple, its analytical representation is also simple. For instance if we assume that the Earth's mass distribution is spherically symmetric, then the magnitude of the force acting on a satellite is

$$F = \frac{Gm_E m_s}{r^2} \ ,$$

where r is the distance between the center of the Earth and the satellite. The position vector from the center of the Earth to the satellite is $\bar{r}$. The gravitational force acting on the satellite points to the Earth; therefore, the previous equation becomes

$$\bar{F} = -\frac{Gm_E m_s}{r^3} \ \bar{r}. \tag{11.1}$$

Note that the minus sign shows that the force is directed to the Earth and $\bar{r}$ is directed away from the Earth.

This vector equation corresponds to three scalar equations, giving the three components of the force field. Let the components of $\bar{r}$ be x, y, and z in a usual rectangular coordinate system centered at the Earth's center and let the components of the force be F_x, F_y and F_z. The vector equation can be written as

$$F_x = - \frac{Gm_E m_s}{r^3} x, \; F_y = - \frac{Gm_E m_s}{r^3} y \; \text{and}$$

$$F_z = - \frac{Gm_E m_s}{r^3} z \, ,$$

where $r = \sqrt{x^2 + y^2 + z^2}$.

The idea of the potential function now enters in the following form. Find a function, depending on the variables x, y, z so that its partial derivatives with respect to x, y, and z give the above three force components. This function is called the potential function.

In analytical form, find the function $U(x, y, z)$ so that

$$F_x = \frac{\partial U}{\partial x} \, , F_y = \frac{\partial U}{\partial y} \; \text{and} \; F_z = \frac{\partial U}{\partial z} \, .$$

If we find such a function, then the three components of the force field can be represented by one scalar function U, instead of three functions, F_x, F_y and F_z. The reader will appreciate that such a delightful simplification does not exist in general, i.e. for any field, and therefore the existence of a potential function is not obvious. Consider for instance the force components in the above gravitational example. If there exists a potential function U, then the relation

$$\frac{\partial^2 U}{\partial x \, \partial y} = \frac{\partial^2 U}{\partial y \, \partial x}$$

must hold. But in terms of the given force components this equation becomes

$$\frac{\partial}{\partial x} F_y = \frac{\partial}{\partial y} F_x \, .$$

This is one of the requirements which the force field must satisfy. For our gravitational field this equation and two similar ones for the x, z and y, z combinations are satisfied. Therefore, the gravitational field has a potential function.

It is at this point that we must conclude our general discussion and omit the methods known for the determination of U for given $\bar{F}$ fields since this problem is not a part of introductory orbit mechanics.

It is, however, simple to verify that the potential function

$$U = \frac{Gm_E m_s}{r} \qquad (11.2)$$

satisfies our requirements and the components of the force field can be obtained as the partial derivatives of U. In vectorial language

$$\bar{F} = \frac{dU}{d\bar{r}} = \nabla U \,,$$

since the gradient of a function U is defined by

$$\nabla U = \frac{\partial U}{\partial x}\,\bar{i} + \frac{\partial U}{\partial y}\,\bar{j} + \frac{\partial U}{\partial z}\,\bar{k} \,,$$

where $\bar{i}, \bar{j}$ and $\bar{k}$ are the unit vectors in the x, y, z directions.

Note that if the relation between the potential function U and the force vector $\bar{F}$ is given by

$$\bar{F} = -\nabla U \,,$$

then our potential function becomes

$$U = -\frac{Gm_E m_s}{r} \,.$$

This double use of minus signs is popular in the classical literature and only Poincaré's authority and encouragement allows us to disregard it.

The concept of potential energy now follows simply by the combination of the equation of motion and the use of the potential function. In an inertial system fixed at the center of the Earth, we have

$$\bar{F} = m_s \ddot{\bar{r}}$$

and

$$\bar{F} = \Delta U = \frac{dU}{d\bar{r}} \,,$$

or

$$m_s \ddot{\bar{r}} = \frac{dU}{d\bar{r}} \,.$$

Multiplying both sides by $\dot{\bar{r}}$ using dot-products, we have

$$m_s \dot{\bar{r}} \cdot \ddot{\bar{r}} = m_s \frac{d}{dt} \frac{(\dot{\bar{r}})^2}{2}$$

and

$$\frac{dU}{d\overline{r}} \cdot \dot{\overline{r}} = \frac{dU}{dt} \ ,$$

or

$$m_s \frac{d}{dt} \frac{(\dot{\overline{r}})^2}{2} = \frac{dU}{dt} \ .$$

From this we have after integration:

$$m_s \frac{(\dot{\overline{r}})^2}{2} = U + constant \ ,$$

or

$$m_s \frac{(\dot{\overline{r}})^2}{2} - G \frac{m_E m_s}{r} = constant \ .$$

This is the classical form of the energy conservation principle, stating that the sum of the kinetic and potential energies is constant. The negative potential energy is the same as the potential function. (Note that if the previously mentioned alternate relation between U and $\overline{F}$ is used - with negative signs - the result will be the same.)

In the above discussion the Earth was not influenced by the gravitational effect of the satellite; therefore, the energy equation (after dividing by m_s and multiplying by 2) becomes

$$v^2 - \frac{2Gm_E}{r} = k \ .$$

The mass m_E is to be replaced by $m_1 + m_2$ if the two-body relative motion equations are used, as mentioned at the end of Chapter 3. In this case the potential function is

$$U = \frac{G(m_1 + m_2)}{r}$$

and the equation of relative motion becomes

$$\ddot{\overline{r}} = \nabla U \ .$$

Non-spheriodicity of the Earth.

The actual gravitational field of the Earth can not be accurately described by the simple potential function mentioned above, which corresponds to a point mass, or to a homogeneous sphere, or to a spherically symmetric mass distribution.

Since the satellite's mass cancels in the equations of motion, the potential function of the Earth, U, in satellite dynamics is usually given per unit mass of the satellite, i.e.

$$U = \frac{Gm_E}{r} \text{ , or } U = \frac{\mu}{r} \text{ ,}$$

if spherical mass distribution is assumed.

Now consider the Earth formed of mass elements Δm_i, each producing a gravitational force on the satellite at a distance r_i and described by the potential function

$$\Delta U_i = \frac{G \Delta m_i}{r_i} \text{ .} \tag{11.3}$$

The total effect is obtained by summation (or by integration) and it becomes

$$U = G \int \frac{dm}{r} \text{ ,} \tag{11.4}$$

where the integration is to be extended to the mass of the Earth or to the central body whose gravitational effect is to be evaluated. If the density is variable, the above integral might be written as

$$U = G \int \frac{\rho dV}{r} \text{ ,}$$

where $\rho = \rho(x, y, z)$ is the density, dV the volume element and the volume integral is again extended for the body of the Earth. If these equations are applied to bodies with spherically symmetric mass distributions, we obtain for the potential, due to M at distance r, the formula

$$U = \frac{GM}{r} \tag{11.5}$$

or

$$U = \frac{\mu}{r} \text{ .}$$

Most of the natural celestial bodies do not possess spherically symmetric mass distribution, and to determine the gravitational force fields (or potential) the previously mentioned volume integral must be evaluated.

The shape of the body is not the only essential property which determines the gravitational potential. The mass distribution or the density variation also must be taken into consideration. If the density of the body is constant, then the shape is the essential factor. Deviation from the spherical shape (for uniform density) may be approximated by ellipsoidal bodies, and this is usually the approach for natural celestial bodies. Since the gravitational potential for ellipsoids (of uniform mass distribution) may be represented by Legendre polynomial expansions, this technique became the most popular approach in describing the gravitational properties of bodies. The approach used in geodesy today is straightforward but not necessarily the final word, and improvements in the basic approach and in the principles used might be forthcoming in the near future. The present method is to write out the potential using Legendre polynomials and from satellite observations evaluate the coefficients in the expansion.

Another approach is known as the inverse problem of celestial mechanics. Usually a force field is given, and for a set of initial conditions we are establishing an orbit, often by numerical integration. This problem in celestial mechanics might be called the direct problem.

If the orbit is given by observations, we might inquire about the force field which produced this orbit. This is known as the inverse problem and it is considerably more complicated than the direct problem, for several reasons. It can be shown for instance that the force field is not uniquely determined from an orbit, or in other words, there are several force fields which can produce the same orbit. If the general functional form of the force field or of the potential is given (such as in the case of the previously mentioned Legendre expansion) then the orbit might be used to determine the coefficients in the expansion. The problem with this "pre-determined potential" approach is that the body, such as the Earth, usually is unaware of our selection of the functional form of its potential, and when the coefficients in our assumed series approximation are evaluated they often show dependence on the orbit. This is especially true for the higher order gravitational coefficients of the usual Legendre series approach.

It might be concluded that the inverse problem of celestial mechanics, often referred to as geodesy, is an unsolved problem and at the present only approximate analytical descriptions are available for the potential functions of natural bodies.

Regarding the shape of the Earth we use the ellipsoidal approximation and then the corresponding infinite series expansion for the gravitational potential. Consider a spherical Earth first with homogeneous mass distribution, i.e. with constant density. An alternate formulation of the basic description is to consider the Earth made of concentric spherical shells of uniform density. For these models as mentioned before, the potential is identical with a point mass potential.

Now consider this spherical Earth and exert radial forces at the North and South poles. Another way to describe this change of shape, consider a small, spherical balloon which is being pushed between our two hands in opposite (radial) direction. The result is an ellipsoid with the equatorial radius ($\overline{CE}$) larger than the polar radius ($\overline{NC}$), as shown in Figure 11.1-(a). This deformation results in an ellipsoid where the line connecting the North and South poles (NS) is an axis of rotational symmetry.

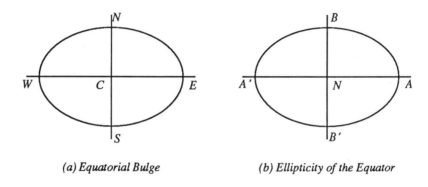

(a) Equatorial Bulge   (b) Ellipticity of the Equator

Figure 11.1. Approximate Shape of the Earth.

The WE axis shown on the figure is also an axis of symmetry since $\overline{WC} = \overline{CE}$. The asymmetry and the ellipticity occur because of the difference between the $\overline{NC}$ and $\overline{CE}$ distances. This deformation and asymmetry is usually referred to as the equatorial bulge. The difference in the axes of the $WNES$ ellipsoid is $\Delta = 21.385\ km$. The oblateness or flattening is described by

$$f = \frac{\overline{CE} - \overline{CN}}{\overline{CE}} = \frac{\Delta}{R_e}.$$

Here $\overline{CE} = 6378,136\ km$ and $\overline{CN} = 6356.751\ km$, giving $f = 1/298.258$.

Note that the corresponding value for Jupiter is $1/16$, making its equatorial bulge about 18 times larger than the Earth's.

The second step in the process of deformation is to destroy the rotational symmetry with respect to the NS polar axis. This might be accomplished by taking our balloon and pushing by our hands at two opposite sides of the equatorial plane as shown in Figure 11.1-(b). The ellipticity of the equatorial plane is represented by

$$\frac{\overline{NA} - \overline{NB}}{\overline{NA}} = \frac{\Delta'}{R_e}$$

which is approximately 10^{-5} corresponding to $\Delta' \approx 210\,ft$.

Up to this point we have discussed asymmetries in the polar and in the equatorial planes still having symmetry with respect to the equatorial plane. The third deformation introduces asymmetry with respect to the equatorial plane which results in a pear-shape configuration.

In the following the potential function will be given due to the above-discussed distortions. It is recalled once again that the gravitational field of the Earth is much more complicated than the above-described elliptic deformations with constant density.

The potential function of a homogeneous ellipsoid corresponding to the equatorial bulge is

$$U = \frac{\mu}{r}\left[1 - \sum_{k=2}^{\infty} \frac{J_k R_e^k}{r^k} P_k(\sin\beta)\right] \tag{11.6}$$

where R_e is the Earth's equatorial radius,

β is the latitude, referred to the equatorial plane,

r is the distance from the center of mass,

P_k is the k-th order Legendre polynomial, and

J_k is called the k-th order zonal harmonics.

The Legendre polynomials are given by various general formulas, the simplest of which is

$$P_k(x) = \frac{1}{2^k k!} \frac{d^k}{dx^k} (x^2 - 1)^k .$$

From this we have

$$P_o(x) = 1, P_1(x) = x, P_2(x) = \frac{1}{2}(3x^2 - 1), P_3(x) = \frac{1}{2}(5x^3 - 3x), etc.$$

If the body is assumed to be symmetric about the equatorial plane, the odd harmonics are zero.

The above formula for the potential follows from Eqn. (11.4) assuming constant density and no ellipticity of the equator. The zonal harmonics, as evaluated from satellite observations are

$$J_2 = 1.0826 \times 10^{-3}; \ J_3 \approx -2 \times 10^{-6}; \ J_5 \approx -2 \times 10^{-7} .$$

For comparison, note that for Jupiter $J_2 \approx 0.015$, corresponding to a much larger flattening than the Earth's, as mentioned before.

If the ellipticity of the equator and the asymmetry with respect to the equatorial plane are also taken into consideration, then in addition to the zonal harmonics, also tesseral and sectorial harmonics enter the picture and the potential becomes

$$U = \frac{\mu}{r} \left[1 - \sum_{k=2}^{\infty} \sum_{j=0}^{k} J_k^{(j)} \left[\frac{R_e}{r} \right]^k (\cos j\lambda)\, P_k^{(j)} (\sin\beta) \right] , \qquad (11.7)$$

where

$$P_k^j(x) = \frac{1}{2^k k!} (1 - x^2)^{j/2} \frac{d^{j+k}}{dx^{j+k}} (x^2 - 1)^k ,$$

and λ is the longitude measured from the major axis of the equatorial ellipse (36° *W*).

The zonal harmonics correspond to $j = 0$, and the previously given formula (Eqn. 11.6) is obtained if we write $J_k^{(0)} = J_k$ and $P_k^{(0)} = P_k$. The tesseral harmonics refer to any value of $j \neq k$. For the sectorial harmonics $j = k$. The zonal harmonics depend only on the latitude, since when $j = 0$, all terms containing λ are missing. The sectorial harmonics depend only on the longitude, and the tesseral harmonics depend on both the latitude and longitude.

The potential function containing the terms which significantly influence the orbit is

$$U = \frac{\mu}{r} \left[1 - \frac{J_2^{(0)}}{2} \left[\frac{R_e}{r} \right]^2 (-1 + 3\sin^2\beta) - 3J_2^{(2)} \left[\frac{R_e}{r} \right]^2 \cos2\lambda\ \cos^2\beta \right] .$$

The "main problem" of satellite dynamics is defined as the problem when only the term with the $J_2^{(0)} = J_2$ coefficient is kept in the above potential function. The corresponding effects are periodic perturbations concerning the orbital elements, with the exception of Ω and ω, which show secular variations, given by

$$\dot{\Omega} = -\frac{3n}{2} J_2 \left[\frac{R_e}{p} \right]^2 \cos i \qquad (11.8)$$

and

$$\dot{\omega} = \frac{3n}{2} J_2 \left[\frac{R_e}{p} \right]^2 \left(2 - \frac{5}{2} \sin^2 i \right) . \qquad (11.9)$$

These equations give the angular variations in rad/sec. If the mean motion n is replaced by 2π in the above equations, we have the angular variations in radians per revolution, since $nP = 2\pi$.

The corresponding results are

$$\Delta\Omega = -3\pi J_2 \left[\frac{R_e}{p}\right]^2 \cos i$$

and

$$\Delta\omega = 3\pi J_2 \left[\frac{R_e}{p}\right]^2 (2 - \frac{5}{2} \sin^2 i).$$

Note that the nodal line usually regresses, since $\dot{\Omega} < 0$ when $0 < i < \pi/2$. The argument of perigee increases or decreases depending on the value of the inclination. The critical angle of inclination is given by

$$\sin i_c = \frac{2}{\sqrt{5}} = 63°26'5.8'' ,$$

which corresponds to $\dot{\omega} = 0$ in Eqn. (11.9).

The computation of $\dot{\omega}$ might be simplified if it is related to $\dot{\Omega}$ since

$$\dot{\omega} = -\frac{4 - 5 \sin^2 i}{2 \cos I} \dot{\Omega}$$

The effects of the sectorial harmonics are important for the stability of geosynchronous satellites since their orbits are in the equatorial plane. Because of the ellipticity of the equator, we have two stable and two unstable positions at the geosynchronous altitude with $i = 0$. The stable points are along the minor axis of the ellipse shown in Fig 11.1-(b) as the line $\overline{BB'}$. The unstable points are along the major axis, i.e. in the $\overline{AA'}$ direction. Station keeping is of considerable importance for communication satellites, and therefore the desirable locations are the stable points where the fuel requirements for orbital stabilization are minimal.

Other Perturbations.

In addition to the non-spheriodicity of the Earth, there are several other effects which result in deviations from two-body orbits of satellites. It is important to note that these effects depend on the orbit; therefore, during the lifetime of a satellite the importance of terms in the equations governing its motion could change in time. It should be realized that depending on the accuracy requirements certain effects can be neglected.

Satellites in equatorial, geosynchronous orbits are primarily affected by the ellipticity of the equator, by their longitude (location) and by solar and lunar perturbations.

If the altitude of the satellite is below 1600 km, the solar and lunar effects (so-called third body effects) might be neglected.

The atmospheric drag becomes important below 600 km, and it strongly depends on the shape and mass of the satellites. It can be shown that atmospheric drag has a secular effect on the semi-major axis, which decreases, and on the eccentricity, which approaches zero. The orbit becomes more and more circular with decreasing semi-major axis until entry occurs. The precise evaluation of drag effects presents one of the most difficult problems in orbit mechanics because of our limited knowledge of the variation of density with location and time. The unpredictable solar activities and their effects on the density make the entry computations rather uncertain.

The atmospheric drag per unit mass of the satellite is usually evaluated from the equation

$$D = \frac{1}{2} c_D \rho V^2 A \; ,$$

where c_D is the drag coefficient, ρ is the density, V the velocity and A is the area of the satellite normal to the velocity vector. The density depends on the location and on the time as mentioned before, but the major effect can be represented by the equation

$$\rho = \rho_0 e^{-\alpha h} \; ,$$

where ρ_o and α are constants and h is the altitude. The area A might also vary since it depends on the attitude of the satellite and, therefore, it can change if the satellite is not spherical. The above formula for the drag might be modified by introducing the ballistic coefficient,

$$B = c_D \frac{A}{m} \; ,$$

where m is the mass of the space probe. The density variation is often represented in tabular form (Jacchia) instead of by the above approximate equation. For all the above reasons, numerical integration becomes mandatory to determine the atmospheric effects on low elevation satellites.

Solar radiation effects are important for low-density (balloon) satellites, with large surface area and with small mass.

The previously mentioned deep-space orbits present major problems in celestial mechanics. These high eccentricity satellites are influenced by atmospheric drag and zonal gravitational harmonics during the time they are near perigee. As these satellites approach their apogee the perturbations due to atmospheric drag and zonal gravitational harmonics become less important than the lunisolar effects.

The effects of the Earth's magnetic field and of collisions with charged and non-charged particles are generally small on the orbital elements.

Collisions with orbital debris, such as parts of inactive satellites, rocket bodies, parts of missiles, etc. present very serious danger to the functioning of satellites, space stations and space vehicles in general. At high altitudes the dispersion of the debris is such that the probability of impact is small. The geosynchronous altitude is an exception since only recently it became mandatory that inactive communication satellites be removed. At low altitudes the probability of impact is reduced since the "space garbage" sooner or later reenters the atmosphere and burns up.

Once again, the inverse problem should be mentioned since satellite orbits offer information, besides higher order gravitational coefficients, about the density of the Earth's atmosphere. Orbital decay allows the determination of the density distribution and its dependence on the altitude.

In conclusion a recently (1974) proposed technique, using tethered satellites is mentioned. This system consists of a space station or a space vehicle such as a space shuttle which, while in orbit, deploys a small satellite into a region where the atmospheric density is to be measured. This satellite will not reenter since it is tethered and from its dynamical behavior the atmospheric density can be determined. Other possible uses of such tethered satellite systems are to generate electricity using the Earth's magnetic field and to allow altitude changes of the shuttle.

The basic references are by Brouwer and Clemence (1961) and by Kovalevsky (1963). Concerning the still intriguing problems of drag and radiation effects see Battin's (1987), Escobal's (1965), Fitzpatrick's (1970) and King-Hele's (1964) books. A recent report on tethered satellites is by V.R. Bond, NASA-JSC-22681, (1987).

EXAMPLES

1. Consider an artificial Earth satellite in a plane at 30° from the Equator ($i = 30°$) with perigee and apogee heights $h_p = 100\ miles$ and $h_a = 520\ miles$. Neglecting the effects of drag, find the secular variations of the orbital elements.

The semi-major axis and eccentricity are related to h_a and h_p by the well known equations:

$$a(1-e) = R_e + h_p = r_p$$

and

$$a(1+e) = R_e + h_a = r_a ,$$

where R_e is the equatorial radius of the Earth. Note that the elevations are denoted by h_p and h_a. The corresponding distances from the center of the Earth are r_p and r_a.

The above equations give

$$a = \frac{r_p + r_a}{2} \quad \text{and} \quad e = \frac{r_a - r_p}{r_a + r_p} ,$$

or

$$e = \frac{r_a - r_p}{2a} .$$

Note that the second equation for e is simpler than the first but it contains a previously computed value (a). Formulas using the original inputs (r_a and r_b) have the advantages of showing the functional dependence of the element (e) on the inputs as well as avoiding propagating errors made in the previously computed element (a), as mentioned before. Also note that the above equations may be written as

$$a = R_e + \frac{h_a + h_p}{2}$$

and

$$e = \frac{h_a - h_p}{2R_e + h_a + h_p} .$$

The semi-latus rectum is given by $p = a(1 - e^2)$, or by

$$p = \frac{2r_a r_p}{r_a + r_p} \ ,$$

and the ratio needed to evaluate $\dot{\Omega}$ and $\dot{\omega}$ is

$$\frac{R_e}{p} = \frac{R_e(r_a + r_p)}{2r_a r_p} \ .$$

The mean motion is computed from Kepler's law:

$$n = \left[\frac{\mu}{a^3} \right]^{1/2} \ ,$$

or

$$n = \left[\frac{8\mu}{(r_p + r_a)^3} \right]^{1/2} \ .$$

The time derivatives of Ω and ω become

$$\dot{\Omega} = -A \ \cos i \ ,$$

and

$$\dot{\omega} = A \left(2 - \frac{5}{2} \ \sin^2 i \right) \ ,$$

where

$$A = \frac{3J_2 R_e^2}{4r_a^2 r_p^2} \ \sqrt{2\mu(r_a + r_p)} \ .$$

Note that these equations allow direct use of the original input values. For the numerical values mentioned above, we have

$a = 6877.034 \ km$,

$e = 0.04914$,

$p = 6860.425 \ km$,

$\dfrac{R_e}{p} = 0.92970$,

$n = 1.10705 \times 10^{-3} \ rad/sec$,

$P = 5675.61 \ sec$,

$\dot{\Omega} = -1.3457 \times 10^{-6} \ rad/sec = -6.66°/day$,

$\dot{\omega} = 2.1366 \times 10^{-6} \ rad/sec = 10.58°/day$.

The changes of Ω and ω in one revolution of the satellite are $\Delta\Omega = -7.64 \times 10^{-3}\,rad$ and $\Delta\omega = 12.13 \times 10^{-3}\,rad$.

2. The artificial satellite Explorer 6, known as 1959 $\delta2$ was discussed in Chapter 6, Example 7. Using the results obtained there, compute the major secular effects on the node and on the argument of perigee when the inclination of the orbit to the equatorial plane is 45°.

The regression of the ascending node becomes $\dot{\Omega} = -4.155 \times 10^{-3}\,rad/day$ or $-0.238°/day$, and the progression of the apsidal line is $\dot{\omega} = 4.407 \times 10^{-3}\,rad/day$ or $0.252°/day$.

3. The semi-major axis and the eccentricity of an artificial Earth satellite are 8676 km and 0.19, and it is in an orbit inclined to the equatorial plane by 34°. The change of the argument of perigee is 4.41°/day and of the nodal line is −3.019°/day. Compare the two values of J_2 which the above observations give.

Solving Eqns. (11.8) and (11.9) for J_2 we have

$$J_2 = - \frac{2\dot{\Omega}}{3n} \left[\frac{p}{R_e} \right]^2 \frac{1}{\cos i} = 1.07916 \times 10^{-3} \,,$$

and

$$J_2 = \frac{4\dot{\omega}}{3n} \left[\frac{p}{R_e} \right]^2 \frac{1}{4 - 5\,\sin^2 i} = 1.07275 \times 10^{-3} \,.$$

Note that the difference between these two computed values is in the 6th decimal and that the presently accepted value of $J_2 = 1.08263 \times 10^{-3}$ shows a 5-figure agreement with our value computed from $\dot{\Omega}$.

4. Find the angles of inclination for which the motions of the nodal line and of the argument of the perigee are in resonance.

By resonance we mean that the ratio of the angular velocities of these motions are rational numbers or

$$-\frac{\dot{\omega}}{\dot{\Omega}} = \frac{n_1}{n_2} \,,$$

where n_1 and n_2 are integers. From the above relation we have

$$n_1 \dot{\Omega} + n_2 \dot{\omega} = 0$$

which is the usual way to express resonance conditions. If the right side of the above equation is only approximately zero, we have near or almost resonant motion. Substituting in the above equation the expressions given by Eqns. (11.8) and (11.9) we have

$$k = \frac{n_1}{n_2} = \frac{5 \cos^2 i - 1}{2 \cos i} \; ,$$

where k is any rational number.

All angles of inclinations which satisfy the above equation result in resonance.

For instance for $k = 1$, we have $i = 46.378°$ or $i = 106.852°$ and for $k = 1/2$ the resonance condition gives $i = 56.065°$ or $i = 110.993°$. (Note that resonance also exists for negative values of k. Regarding $k = 2$, which value gives $i = 0$, see Chapter 9, where ω was introduced as the proper variable for this special case.)

PROBLEMS

1. Jupiter's fifth satellite Amalthea shows a precession of the line of apsides, $\dot{\omega} = 2.51 \; degrees$ per day. The orbit is approximately circular with $a = 181,200 \; km$ and its inclination may be neglected. Find the oblateness parameter (J_2) of Jupiter.

2. Find the oblateness parameter (J_2) of a planet having the same radius and mass as the Earth by observing the orbit of an artificial satellite around the planet having pericenter altitude, $h_P = 200 \; km$; apocenter altitude $h_A = 400 \; km$; inclination from the equatorial plane, $i = 30°$; and pericenter shift, $\Delta \omega = 10°/day$.

Chapter 12. The Problem of Three Bodies

Introduction.

One of the characteristic and interesting observations we can make concerning various aspects of the problem of three bodies is that it attracted the attention of many of the outstanding contributors in celestial mechanics, probably because of the inherent and basic difficulties involved. A short summary of the basic contributions follows.

Newton was one of the first to investigate the problem of three bodies formed by the Sun, the Earth and the Moon. He complained that the problem gave him headaches and kept him awake, but finally he was still able to compute the motion of the perigee of the lunar orbit within 8% of the observed value (1687).

Euler in 1760 proposed the highly special problem of three bodies known as the problem of two fixed force-centers which is solvable by elliptic functions and which formed the basis of Vinti's satellite theory in 1961. In this problem two of the bodies are fixed and the third moves in their gravitational field. Euler also proposed in 1772 the use of a rotating or synodic coordinate system for the restricted problem of three bodies, which allowed Jacobi in 1836 to introduce the integral of the motion named after him. This problem will be discussed in detail since it is one of the important dynamical systems in orbit mechanics.

Lagrange, also in 1772, has shown the existence of equilibrium points in the restricted problem, which led 134 years later to the discovery of the Trojan asteroids.

The Sun-Earth-Moon three-body problem known as the lunar theory in the form still being used today is the result of Hill's (1878) and Brown's (1896) contributions.

Poincaré, concentrating on the qualitative aspects of the problem, showed its non-integrability in 1899 and treated the theory of periodic orbits.

K.F. Sundman in 1912 offered analytical regularization, and E. Stromgren and his Copenhagen school computed an impressive number of families of periodic orbits between 1913 and 1939. Sir George Howard Darwin (fifth son of Charles Robert Darwin) studied the evolution of periodic orbits for more practical cases than Stromgren and published his results between 1897 and 1911. Darwin and Stromgren concentrated on the restricted version of the problem of three bodies.

The dynamical system known as the restricted problem of three bodies may be obtained physically or derived mathematically from the general problem of three bodies. Therefore, a few words should be said about the general problem which forms the basis of the restricted problem. The space dynamics applications of the general problem belong to the distant future while the restricted problem is the basic model for present day lunar and interplanetary trajectories as well as for high eccentricity, large semi-major axis satellite orbits. In astronomy, especially in stellar dynamics, the general problem of three bodies is dominant.

The basic and important difference between the problems of two and three bodies is that the former is integrable but the latter is not. The somewhat complicated and not uniformly accepted concept of integrability is associated with Bruns (1887) and Poincaré (1890), and it usually refers to the availability of generally valid analytical solutions. For any given set of initial conditions we can predict the quantitative and qualitative behavior of the problem of two bodies. This is not the case for the problem of three bodies. If an analytical solution is desired for a certain set of initial conditions for the problem of three bodies, this solution is usually expressed by infinite series. If the series converge for any length of time we have an analytic solution for a given set of initial conditions. Changing the initial conditions, the solution might change completely, and the series might diverge. The solution, consequently, has no general validity. Furthermore, solutions given by infinite series (even when they converge) do not offer a qualitative picture. For the problem of two bodies, the initial conditions will tell us the qualitative nature of the solution, such as the type of conic section which the orbit will follow. This is not the case of the problem of three bodies, where the qualitative properties of the solution are generally not known.

Another representation of the concept of integrability is that the system has a sufficient number of generally valid and independent invariant relations (integrals) between the variables, so that analytical solutions can be obtained. The idea of integrability is not a simple one and it does not belong to an introductory textbook; nevertheless, its importance warranted the above short description.

As an additional note the reader is reminded of two, already mentioned concepts of some interest. First, it is recalled that the time dependence of the variables in the solution of the problem of two bodies cannot be expressed by closed form functions, as made clear by Kepler's equation. This fact does not mean that this problem is not integrable. The second remark is related to the solvability of a dynamical problem versus its integrability. By solution we can mean a special solution obtained by numerical integration, valid for a certain given time. Such solutions, of course, exist and might be obtained for the problem of three bodies. As the time increases the special solution will lose its accuracy and validity. In general as $t \rightarrow \infty$ the numerical solution becomes meaningless. Not so for the problem of two bodies. If the initial conditions indicate an elliptic solution, the participating two bodies will remain on their elliptic orbits as $t \rightarrow \infty$.

The General Problem of Three Bodies.

After these introductory remarks we are now ready to discuss the basic ideas of the general problem and some of the practical details of the restricted problem of three bodies.

The basic and original definition of the problem of three bodies is as follows: three point masses (or bodies of spherical symmetry) gravitationally attract each other; for a given set of initial conditions, find the resulting motion. The motion in general takes place in three dimensions, and there are no restrictions concerning the masses, the initial positions and the initial velocities. Without actually writing down the equations of motion we can expect three second order differential equations for the three position vectors of the three bodies, forming a $2 \times 3 \times 3 = 18$-th order system. The energy is conserved since it is a conservative system; the angular momentum is conserved since there are no moments acting; the center of mass of the three bodies moves with constant velocity; but there are still not enough integrals to "solve" the problem, as mentioned before.

Ambitious readers will find little difficulty and considerable satisfaction by writing down the equations of motion for the general problem of three bodies and comparing their results with any of the standard advanced references mentioned in the Appendix. Once the equations are available they will not be able to resist the temptation to derive the integrals of energy, angular momentum, and center of mass. The method is the same as used in Chapter 3 in connection with the problem of two bodies.

The Restricted Problem of Three Bodies.

There are several modifications and special cases of the general problem of three bodies, one of which is of considerable importance in space research. This is the restricted problem mentioned before. Two of the three bodies have much larger masses than the third. As a result, the motion of the two larger masses will not be influenced by the third body, but the larger bodies will govern the motion of the small body. A good example of immediate practical importance is the system consisting of the Earth, the Moon and a space probe traveling on a lunar trajectory. The masses of the participating bodies are related by

$$m_E : m_M : m_P \approx 100 : 1 : 10^{-19}$$

where m_E, m_M and m_P are the masses of the Earth, of the Moon and of a 6000 kg probe. If the forces are computed we might see that the effect of the probe on the Earth is always 16 orders of magnitude smaller than the Moon's effect on the Earth. Furthermore, the effect of the Earth on the Moon is always 16 orders of magnitude larger than the effect of the probe on the Moon. We can continue these relative force evaluations (using only two-body effect) and find the Earth's and Moon's effects on the probe as it travels from the Earth to the Moon. When it is in the vicinity of the Earth, the Moon's effect may be neglected since it is 10^{-5} times smaller than the Earth's. When the probe is close to the Moon, the Earth's effect is 10^{-3} times smaller than the Moon's effect.

If the model of the restricted problem is applied to a very large "probe" such as to the Moon as influenced by the Earth and by the Sun, the above 10^{-16} neglected order of magnitude becomes 10^{-2}. Therefore, the restricted problem should be used only as a first approximation to establish the orbit of the Moon (see the examples at the end of this chapter).

The essential idea of the restricted problem is that we can separate the motion of the two large bodies (often called the primaries) and solve this two-body problem first without considering the third small body. After the problem of two bodies is solved we investigate the motion of the small body in the (known) gravitational field of the two large bodies. In many problems of interest in space dynamics the primaries move on approximate circles (Earth and Moon, Sun and Jupiter, etc). This results in the simplest form of the problem, known as the circular restricted problem of three bodies. For this case we have the Jacobian integral which gives an analytical relation between the probe's position and velocity. Note that this relation is not the "solution" of this, still nonintegrable dynamical system.

The applications of the restricted problem are of considerable interest in space dynamics. In the following, two of these are discussed in detail: the equilibrium points and the curves of zero velocity.

When the primaries move on circles, there are five points in the plane of their motion where the forces acting on a probe are balanced. These forces are the gravitational attractions of the large masses on the probe and the centrifugal force acting on the probe revolving with the primaries. This revolving system is known as the synodic system in which the primaries are fixed. In a fixed inertial system, called the sidereal system, the primaries are moving in circles. Three of the equilibrium points are located on the line connecting the primaries and two points form equilateral triangles with the primaries. The three collinear points are unstable and the two triangular points are stable for small mass ratios. This means that if outside forces are acting on the probe which is placed at any of the collinear equilibrium points, the probe will depart. A probe placed at the triangular points will librate instead of escaping, provided the mass ratio of the primaries is smaller than 0.0385. The triangular points, therefore, are also called points of libration. (Libration is an oscillatory motion around an equilibrium point.) Because of their discoverer, the equilibrium points are also known as Lagrangian points. The condition of the stability of the triangular points (i.e. the small mass ratio of the primaries) is satisfied for systems of interest in space dynamics. For instance for the Earth-Moon system the value of the mass parameter is 0.012. The instability of the collinear locations might be counteracted by station keeping propulsion systems. Figure 12.1 shows the locations of the five equilibrium points for the Earth-Moon system.

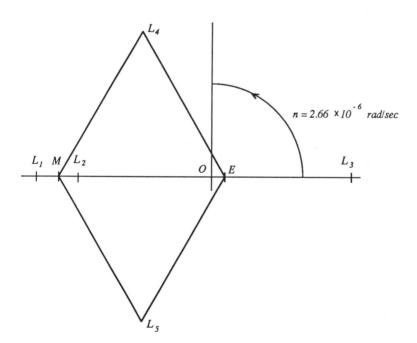

Figure 12.1. Location of the Equilibrium Points in the Restricted Problem of Three Bodies.

Some of the numerical values describing this system are

$$m_E = 5.97 \times 10^{24} \, kg$$

$$m_M = 7.35 \times 10^{22} kg$$

$$\overline{EM} = 384.4 \times 10^3 \, km$$

$$\overline{OE} \approx 0.12 \times \overline{EM}$$

$$\overline{ML_1} \approx \overline{ML_2} \approx 0.16 \times \overline{EM}$$

$$\overline{EL_3} \approx \overline{EM} .$$

The first and second collinear libration points (L_1 and L_2) are located at 16% of the Earth-Moon distance and the third (L_3) is symmetrical to the Moon's location. For the triangular points we have

$$\overline{EL_4} = \overline{ML_4} = \overline{EL_5} = \overline{ML_5} = \overline{EM} .$$

The idea of the curves of zero velocity will be shortly described as the last remark of this chapter. This was originated by Hill in 1878 while establishing his lunar theory. The idea is similar to the use of the conservation of the energy principle which we have seen in connection with the problem of two bodies in the form

$$v^2 = \frac{2\mu}{r} + k.$$

For a given distance (r) and velocity (v), the constant of energy (k) can be computed. The value of this constant cannot change during the trajectory unless the energy of the system is changed by some outside influence or by applying propulsion.

To explain the principle of zero velocity curves, we rewrite the above energy equation in a simplified form:

$$v^2 = \frac{2}{r} + k,$$

where $\mu = 1$, using proper units.

Let us start the motion with $v = 1$ on the unit circle ($r = 1$). The constant of energy becomes $k = 1 - 2 = -1$. During the motion, the velocity-distance relation will always have to satisfy the equation

$$v^2 = \frac{2}{r} - 1.$$

The left side of this equation is zero or positive since it is the square of a real number; therefore, the right side of the equation also must be positive, or zero:

$$\frac{2}{r} - 1 \geq 0.$$

This means that $r \leq 2$ during the motion, or that the body must stay inside of a circle of radius 2. If the body touches the circle its velocity will be zero. The information that the motion must take place inside the $r = 2$ circular region does not tell us anything about the details of the orbit except it offers the qualitative result, according to which the motion cannot leave the region of the zero velocity curve, which in our case is a circle.

For the problem of two bodies this information might be of relatively little interest since all details of the trajectory can be established once the location of the starting point and the components of the initial starting velocity are given. This is not the case when we deal with a non-integrable problem like the restricted problem of three bodies. There the details of the trajectory cannot be established analytically (or numerically) for an arbitrary length of time; therefore, the establishment of regions of possible motion can be of great importance. The equation connecting the velocity of a probe with its position is much more complicated when we deal with the restricted problem of three bodies, for instance with an Earth to Moon trajectory, but it still can be used to establish regions of possible motions. These curves of zero velocity can be obtained from the previously mentioned Jacobian integral, but the subject is above our introductory text. To capture the reader's interest it should be mentioned that Hill used these curves of zero velocity to show that the Moon is "permanently" attached to the Earth. This result is impressive since the general solution of the lunar problem for arbitrary length of time does not exist.

The basic references to curves of zero velocity are by G.W. Hill, "Researchers in the Lunar Theory," American Journal of Mathematics, Vol. 1, Number 5, pp. 129 and 245, 1878 and Szebehely's book (1967), Chapter 4, see Appendix.

The recommended references concerning integrability are Whittaker's Chapter 13 and Wintner's Chapter 3, pp. 194-205, see Appendix.

The effects of the gravitational properties of the primaries and of the solar radiation pressure on the motion around the equilibrium points in the restricted problem of three bodies are discussed in several papers by K.B. Bhatnagar and his school at the University of Delhi (*Space Dynamics and Celestial Mechanics*, D. Reidel Publ. Co., Dordrecht, Holland 1986).

EXAMPLES

1. In this example we will compute the forces acting between the bodies participating in the problem of three bodies formed by the Sun, the Earth and the Moon.

    The force between the Sun and the Earth is

    $$F_{SE} = \frac{GM_S M_E}{d_{SE}^2} = 3.54 \times 10^{19} \, kg.$$

    The force between the Earth and the Moon is

    $$F_{EM} = \frac{GM_E M_M}{d_{EM}^2} = 1.98 \times 10^{17} \, kg.$$

    The force between the Sun and the Moon is

    $$F_{SM} = \frac{GM_S M_M}{d_{SM}^2} = 4.36 \times 10^{17} \, kg.$$

    The following conclusions can be reached:

    a. The Moon's motion is governed by the Sun, rather than by the Earth since

    $$F_{SM} = 2.2 F_{EM} \ .$$

    b. The Earth's motion is governed by the Sun, rather than by the Moon since

    $$F_{SE} = 178.8 F_{EM} \ .$$

    c. The Sun's motion is more influenced by the Earth than by the Moon since

    $$F_{SE} = 81.2 F_{SM} .$$

    d. The motion of the Earth-Sun system can be computed as a problem of two bodies since the effects of the Moon on the Sun and on the Earth are two orders of magnitudes smaller than the force between the Sun and the Earth. This is the reason why the model of the restricted problem of three bodies is used only as a first approximation to compute the orbit of the Moon.

        Note that when the forces were computed the eccentricities of the orbits were neglected and the approximations of $d_{SM} = d_{SE}$ was used. This relation introduces an error of 0.5 percent because of the variable distance between the Sun and the Moon. The effect on the force computation is one percent.

2. Find the location of the first collinear equilibrium point (L_1) in the Earth-Moon system, using the notation of Figure 12.1.

The centrifugal force per unit mass acting on a space probe located at L_1, is

$$F_c = (\overline{L_1 O})n^2 .$$

The Earth's gravitational effect per unit mass on the probe is

$$F_E = \frac{GM_E}{(\overline{EL_1})^2} ,$$

and the Moon's gravitational effect is

$$F_M = \frac{GM_M}{(\overline{ML_1})^2} .$$

The forces are balanced when

$$F_c = F_E + F_M.$$

The sum of the masses of the Earth and of the Moon, $M_E + M_M$, will be denoted by M_T. Kepler's third law gives

$$n^2 = \frac{M_T G}{(\overline{EM})^3} .$$

Since the center of mass is at point O,

$$\overline{OM} = \frac{\overline{EM}}{M_T} M_E ,$$

and since $\overline{L_1 O} = \overline{OM} + \overline{ML_1}$, we have

$$\overline{L_1 O} = \frac{\overline{EM}}{M_T} M_E + \overline{ML_1} .$$

The distance between the Earth and L_1 is

$$\overline{EL_1} = \overline{EM} + \overline{ML_1} .$$

Now we substitute the above expressions for $\overline{L_1 O}$, n^2 and $\overline{EL_1}$ into the force-balance equation and write x for $\overline{ML_1}$, l for $\overline{EM}$, a for M_E / M_T and b for M_M / M_T . The result is

$$x^2 (la + x)(l + x)^2 - l^3 a x^2 - l^3 b(l + x)^2 = 0, \text{ where } a + b = 1 .$$

The unknown in this fifth order equation is the distance $x = \overline{ML_1}$ which gives the location of the first collinear equilibrium point. All the other quantities in this equation are known. The computation of the only positive root by iteration is left to the reader.

The approximate solution is given by

$$x \approx \left[\frac{b}{3(1-b)} \right]^{1/3} l \,.$$

The solution is $x = 61500$ *km*, which distance corresponds to approximately 1.54 times the circumference of the Earth.

PROBLEMS

1. Locate the coordinates of all five equilibrium points for the Earth-Moon system and for the Sun-Jupiter system (and in this way become a scientific descendant of Lagrange).

Concluding Remarks

If readers enjoyed the book, became infatuated with the spirit of our ancestors and wish to contribute to our difficult, but basically simple field, no time was wasted. If they were frightened and overwhelmed by the difficulties, they should realize that their unlimited mental capacity is being challenged. They must remember that there is no gain without pain and joining the club of our giants requires paying the dues.

The beauty of the panorama of celestial mechanics shall never be diminished, since Poincaré's dictum of non-integrability of Newtonian problems above the level of two bodies will not be changed, no matter what magic advanced mathematics, topology or high speed computers will come up with.

Our field is difficult, important and is here to stay to offer challenges to those of us who do not wish to study linear systems but prefer Newton's inverse square law.

The limits of predictability lead to our humbleness and to Sir Isaac Newton. The emergence of instabilities leads us to chaotic motion which penetrates the citadel of deterministic celestial mechanics.

The purpose of this book is to prepare the reader to any assignment, be it along the lines of theoretical astronomy, interplanetary mission analysis, determination of the gravity fields of planets, precise orbit predictions of natural and artificial bodies, or the design of rendezvous in space. The basic principles to the solution of these problems are in the book, together with references to the pertinent literature.

The true excitement of a field of science and engineering comes from unsolved or unsolvable problems. In this respect our field is far superior to others. Applications of our techniques to other fields, such as the instability and singularity present at the origin of the universe, the behavior of elementary particles in supercolliders, guidance and control of space probes and missiles, etc. offer many surprises.

In summary, what the readers have learned was a set of basic principles, applicable to a great variety of fields. But they also learned humility, originating from the fundamental limitations of our field.

Appendix I **Glossary**

Anomalies: the angles describing the motion of a body in a reference frame as independent variables (see eccentric, mean and true anomalies).

Anomalistic year: the mean interval between successive perihelion passages of the Earth.

Aphelion: the point of a planetary orbit which is at the greatest distance from the Sun.

Apoapsis: the point of an elliptic orbit farthest from the focus.

Apogee: the point of an elliptic satellite orbit farthest from the Earth.

Apsidal line: the line connecting the periapsis with the apoapsis.

Argument of latitude: the angle from the ascending node to the location of the body (measured in the orbital plane); also the sum of the argument of periapsis and the true anomaly.

Argument of periapsis: the angle between the nodes and the apsidal line in the plane of the orbit (see Fig. 9.1).

Ascending node: the point in the equatorial plane, or in general, in the reference plane, where the body passes from the Southern to the Northern Hemisphere (see Fig. 9.1).

Astrodynamics: branch of space engineering or astronautics dealing with the orbital and rigid body motion of artificial bodies in space.

Astronautics: branch of engineering dealing with space missions.

Barycenter: the center of mass of a system of bodies.

Celestial latitude: the angle between the ecliptic and a given point, measured along the great circle.

Celestial mechanics: branch of dynamical astronomy, dealing mostly with the motion and dynamics of bodies of the solar system.

Circular restricted problem of three bodies: two bodies with large masses move on circular orbits and influence the motion of a third body with much smaller mass.

Circularization: the change of an elliptic orbit to a circular orbit by change of the velocity.

Dynamical astronomy: branch of astronomy dealing with the motion and dynamics of celestial bodies. It includes celestial mechanics, stellar dynamics, motion of binary stars, positional astronomy, etc.

Eccentric anomaly: the angle at the center of an elliptic orbit, formed by the apsidal line and the radius vector drawn from the center to the point on the circumscribing auxilliary circle from which a perpendicular to the apsidal line will intersect the orbit (see the angle E on Fig. 5.2).

Eccentricity: the distance from the center to the focus divided by the length the semi-major axis.

Ecliptic: the mean plane of the Earth's orbit around the Sun.

Ephemeris: the tabular representation of the position as a function of time of natural or artificial bodies.

Equatorial bulge: see flattening.

Equatorial satellite: a satellite orbiting in the equatorial plane of the Earth.

Equinox: the intersection of the equatorial plane and the plane of the ecliptic.

Escape velocity: the velocity which results in a two-body orbit with zero velocity at infinity.

Flattening: a measure of deviation from a spherical shape, $f = (a - b)/a$, where a is the equatorial radius and b is the polar radius. Also known as oblateness or equatorial bulge, and it is applied to a body generated by the rotation of an ellipse about its minor axis.

General perturbation method: analytical solution of the differential equations describing a perturbed orbit.

Geocentric: referred to the center of the Earth.

Geocentric gravitational constant: the product of the mass of the Earth and the constant of gravity.

Geoid: an equipotential surface of the Earth corresponding to the mean sea level of the open ocean.

Geostationary satellite: a satellite at geosynchronous altitude.

Geosynchronous altitude: the elevation above the Earth's equator, where a satellite's position is fixed relative to the Earth's rotation.

Gravitational constant: the factor of proportionality in Newton's law of gravity.

Gravitational harmonics: the terms in the Legendre series expansion of the gravitational potential.

Gravitational potential: a function, derivative of which gives the gravitational force.

Heliocentric: referred to the center of the Sun.

Heliocentric gravitational constant: the product of the mass of the Sun and the constant of gravity.

Hohmann orbit: the tangential elliptic transfer orbit between two orbits of different radii or semi-major axes (see Fig. 6.2).

Hyperbolic excess velocity: the velocity above escape velocity.

Inclination: the angle between the orbital plane and the reference plane, which is the equatorial plane for planetary satellites and the ecliptic for heliocentric orbits.

Intermediate orbit: an approximation to the actual perturbed orbit (special case is the osculating orbit).

Intermediary orbit: see intermediate orbit.

International ellipsoid: ellipsoid approximating the shape of the Earth.

Invariable plane: the plane containing the center of mass of the solar system and perpendicular to the angular momentum vector of the solar system.

Isochronous stability: kinematical behavior of the disturbed path (Lyapunov's stability).

Kepler's equations: the transcendental relation between mean and eccentric anomaly.

Kepler's laws: three principles describing the motion of the planets in the solar system, generally applicable to the problems of two bodies.

Lagrangian solutions: the equilibrium solutions of the restricted problem of three bodies (see Fig. 12.1).

Lambert's theorem: a relation showing the ellapsed time on a two-body orbit as a function of the chord, of the sum of the radial distances and of the semi-major axis.

Laplace's invariable plane: see invariable plane.

Latitude: the angle between the ecliptic (or the equatorial plane) and the position vector measured at the center of the Sun (or the Earth) along the great circle (or the meridian).

Libration: oscillation about equilibrium points; for instance the variation of the orientation of the Moon with respect to the Earth; also motion of bodies around the triangular equilibrium points in the restricted problem of three bodies.

Linear stability: effect of small disturbances when applied to the linearized equations of motion.

Line of nodes: the line connecting the ascending and descending nodes (see Fig. 9.1).

Longitude: the angle between the vernal equinox and the great circle, measured in the ecliptic or the angle between the Greenwich meridian and the meridian of a given position measured in the equatorial plane.

Longitude of periapsis: the sum of the angles of the longitude of the ascending node and the argument of periapsis (Fig. 9.1).

Lunar theory: a prediction of the motion of the Moon, usually analytical.

Mean anomaly: the product of the mean motion and the interval of time since pericenter passage.

Mean motion: the value of a constant angular velocity required for a body to complete one revolution.

Meridian: the great circle between the North and South poles (terrestrial and celestial) which passes through the point directly above the observer.

Newton's law of gravitation: the gravitational force between bodies is directly proportional to the product of their masses and inversely proportional to the square of their distances.

Nutation: the short-period oscillation of the pole.

Oblateness: see flattening.

Obliquity: the angle between the equatorial and orbital planes.

Obliquity of the ecliptic: the angle between the equatorial and the ecliptic planes.

Orbit: usually the path of a body with respect to another such as planetary orbit around the Sun.

Orbital stability: geometrical behavior of the disturbed orbit (Poincaré's stability).

Orbit mechanics: branch of mechanics dealing mostly with the orbital motion of natural and artificial bodies in space.

Osculating elements: the elements of the perturbed two-body orbit which would be applicable if the perturbation would be eliminated.

Osculating orbit: the two-body orbit which would be followed if the perturbations would be turned off (see also intermediate orbit).

Parabolic velocity: see escape velocity.

Periapsis: the point of an orbit closest to the focus.

Perigee: the point of a satellite orbit closest to the Earth.

Perihelion: the point of a planetary orbit closest to the Sun.

Period: the interval of time between consecutive occasions on which the system is represented by the same vector in the phase space (same position and velocity).

Periodic motion: the motion repeats itself in equal intervals of time (recurrence of dynamical properties).

Periodic perturbations: the periodically changing effect of perturbations.

Perturbations: forces which result in deviations between the actual orbit and reference orbit, such as a two-body orbit. (See also special, general, secular and periodic perturbations.)

Phase space: combination of position and velocity coordinates (six-dimensional for the three-dimensional motion of a point mass).

Planetary theory: prediction of the motion of planets of the solar system, usually analytical.

Polar orbit: it passes over the North and South pole, its inclination is 90 degrees.

Problem of three bodies (general): the dynamics of three gravitationally interacting point masses.

Problem of three bodies (restricted): the modification of the general problem in the case when one of the three bodies, because of its small mass, is not influencing the motions of the other two bodies with much larger masses.

Problem of two bodies: the dynamics of two gravitationally interacting point masses.

Precession: the secular motion of the pole.

Rectilinear orbit: a straight-line orbit, for two bodies with unit eccentricity and zero semi-latus rectum.

Regularization: the elimination of singularities from the equations of motion.

Restricted problem of two bodies: the mass of one of the bodies is much smaller than the mass of the other and the gravitational effect of the smaller body is neglected.

Sectorial harmonics: the term in the Legendre series expansion of the gravitational potential which depend only on the longitude.

Secular perturbation: the continuously increasing effect of perturbations.

Selenocentric: referred to the center of the Moon.

Semi-latus rectum: the distance between the focus and a point on the conic section measured in the direction normal to the apsidal line.

Semi-major axis: the distance from the center of an ellipse or hyperbola to an apsis.

Semi-minor axis: the distance from the center of the ellipse along the line perpendicular to the apsidal line.

Sidereal time: a measure of the Earth's rotation with respect to the stars.

Sidereal year: the time for the Earth to complete one revolution on its orbit with respect to a fixed vernal equinox or with respect to the background stars.

Singularity: the appearance of zero distances between participating bodies, or in general, zero denominators appearing in the equations of motions.

Small divisor: the denominator appearing in perturbation analysis which approaches zero, usually because of resonance conditions.

Special perturbation method: numerical integration of the differential equations describing a perturbed orbit.

Stability: behavior of a dynamical system when disturbances are applied.

Stable motion: the effect of initially small disturbances stay below a given limit.

Stellar dynamics: branch of dynamical astronomy dealing with the motion and dynamics of stars and stellar systems such as clusters and galaxies.

Synodic period of planetary motion: the time between two successive conjunctions of two planets, as observed from the Sun.

Synodic period of satellite motion: the time between two successive conjunctions of the satellite with the Sun, as observed from the satellite's planet.

Synodic system: in general, a coordinate system rotating around the center of mass of the participating bodies.

Terrestrial harmonics: terms in the Legendre series expansion of the Earth's gravitational potential.

Terrestrial latitude: the angle between the equatorial plane and a given point, measured along the meridian.

Tesseral harmonics: the term in the Legendre series expansion of the gravitational potential which depend on the latitude and longitude $(J_k^j, j \neq k)$.

Time of perigee passage: the time when a satellite passes closest to the Earth.

Topocentric: referred to a point on the surface of the Earth.

Trajectory: usually the part of an orbit such as a missile's or rocket's path.

True anomaly: the angle at the focus (nearest the pericenter) between the apsidal line and the radius vector (drawn from the focus to the orbiting body). (See angle f in Figs. 4.1 and 5.2.)

Unstable motion: the effect of initially small disturbances increase above a given limit.

Vernal equinox: the direction where the Sun crosses the equatorial plane from South to North in its apparent motion along the ecliptic (its apparent longitude is zero); the ascending node of the ecliptic on the equatorial plane.

Zonal harmonics: the terms in the Legendre series expansion of the gravitational potential which depend only on the latitude.

Appendix II **Physical Constants**

The reader is reminded that the following constants change as better measurements and observations become available, but their value will never be known "exactly." Most of the numbers presented are based on the 1988 edition of the Astronomical Almanac issued by the Nautical Almanac Office, U.S. Naval Observatory and published by the U.S. Government Printing Office, Washington, D.C. These numbers do not always agree completely with those given by other sources, such as the International Astronomical Union, International Association of Geodesy, etc. For instance the IAG value for the Earth's equatorial radius is 6378136 ± 1 meter, which constant in our table is 6378.14 km.

Some of the entries are redundant since the reader certainly can compute the constant of gravity (G) if the geocentric gravitational constant (GM_E) and the value of the Earth's mass (M_E) are given. The reason for furnishing such redundant values is to facilitate the solution of actual problems in space dynamics. Some of the constants for the Earth are given in the table (rounded off values) and also are listed with the presently existing highest accuracy, showing the error limits.

The effects of uncertainties in orbit mechanics are discussed by G.B. Westrom's article which appeared in the proceedings of a symposium on "Space Trajectories" published by Academic Press, New York (1960) and by a fascinating paper by Sir James Lighthill, "The Recently Recognized Failure of Predictability in Newtonian Dynamics," Proc. Roy. Soc. of London, Vol. A407, pp. 35-50, 1986.

For additional information see the above-mentioned Astronomical Almanac or its Explanatory Supplement (1961, revision published in 1988). A.E. Roy's *Orbital Motion*, listed in the references, also has many of the useful constants.

Table of Constants of the Planets

Planet	R_e (km)	M $(10^{24}kg)$	J_2 (10^{-3})	i (degrees)	a (A.U.)	e
Mercury	2439	0.33022	-	7.006	0.3871	0.2056
Venus	6052	4.8690	0.027	3.395	0.7233	0.0067
Earth	6378.14	5.9742	1.028263	0.001	1.0000	0.0167
Mars	3393.3	0.64191	1.964	1.851	1.5237	0.0933
Jupiter	71398	1898.8	14.75	1.305	5.2030	0.0482
Saturn	60000	568.50	16.45	2.486	9.5281	0.0542
Uranus	25400	86.625	12	0.663	19.1829	0.0459
Neptune	24300	102.78	4	1.769	30.0796	0.0101
Pluto	1500	0.015	-	17.142	39.3396	0.2462

Notes

R_e : equatorial radius

M : mass

J_2 : second zonal gravitational harmonic

i : inclination to the mean ecliptic

a : semi-major axis

e : eccentricity

The heliocentric osculating orbital elements given in this table are referred to the mean ecliptic and equinox of J2000 and are given for February 9, 1988.

Additional Constants of the Solar System

Equatorial radius of the Sun: 696000 *km*

Mass of the Sun: $1.9891 \times 10^{30} kg$

Heliocentric gravitational constant: $GM_s = 1.3271244 \times 10^{20} m^3/sec^2$

Astronomical Unit: $1.49597870 \times 10^8 km$

$$\frac{Mass\ of\ Sun}{Mass\ of\ Jupiter} = 1047.355$$

$$\frac{Mass\ of\ Sun}{Mass\ of\ Earth} = 332946$$

$$\frac{Mass\ of\ Sun}{Mass\ of\ Earth + Moon} = 328900.5$$

Equatorial radius of the Earth: $R_e = (6378136 \pm 1)m$

Polar radius of the Earth: $R_p = (6356751 \pm 1)m$

Flattening of the Earth: $f = 1/298.25781$

Second zonal harmonic of the Earth: $J_2 = (1082626 \pm 2) \times 10^{-9}$

Period of the Earth's revolution (one sidereal year): 356.25636 *days*

Anomalistic year: 365.25964 *days*

Angular velocity of the Earth's rotation:
$$\omega = (7.292115 \pm 10^{-7}) \times 10^{-5} rad/sec$$

Period of the Earth's rotation (one sidereal day):
$$T = 23\ hours\ 56\ min\ 4.1sec$$

Obliquity of the Ecliptic: 23°26′34″ 37″ for 1988.

Geocentric gravitational constant:
$$GM_E = (3986004.40 \pm 0.03) \times 10^8 m^3/sec^2$$

$$\frac{Mass\ of\ Moon}{Mass\ of\ Earth} = 0.0123000$$

Mean radius of the Moon: 1738*km*

Mass of the Moon: $7.3483 \times 10^{22} kg$

Semi-major axis of the lunar orbit: 384400*km*

Moon's orbital period (sidereal): 27.321661 *days*

Eccentricity of the lunar orbit: 0.0549

Constant of gravitation: $G = 6.672 \times 10^{-11} m^3 kg^{-1} sec^{-2}$

Speed of light: 299792458 $m\ sec^{-1}$

Appendix III Annotated List of Major Reference Books

The preparation of such lists presents many difficult choices. Inclusion of too many items will not help readers, but anything important left out will be to their definite disadvantage. The notes added might help the readers to decide how to spend their time and where to turn to satisfy their curiosity. A few books emphasizing the historical aspect are also included to allow the readers to balance science and humanities.

E.N. da C. Andrade, *Sir Isaac Newton*. Macmillan, New York, 1954.

Combination of Newton's life, works, personality, human dimension and detailed biography.

V.I. Arnold, *Mathematical Methods of Classical Mechanics*. Nauka, Moscow, 1974. Translation published by Springer-Verlag, New York, 1978.

Modern mathematical approach to dynamics, including Newtonian, Lagrangian and Hamiltonian formulations. Classical mechanics is related to areas of mathematics such as Riemannian geometry, Kolmogorov's theorem, Lie groups, mapping theorems, etc. This is an advanced text book used by Arnold teaching classical mechanics at Moscow State University.

R.M.L. Baker and M.W. Makemson, *An Introduction to Astrodynamics*. Academic Press, New York, 1960. Second revised edition 1967.

Easy to read, written in clear style, mostly for engineers. Many useful exercises concerning space dynamics.

R.R. Bate, D.D. Mueller and J.E. White, *Fundamentals of Astrodynamics*. Dover, New York, 1971.

This textbook is for introductory engineering courses, emphasizing the systems engineering approach, the practical aspects of orbit determination, lunar and interplanetary trajectories (with patched conic approximations) and ballistic missile trajectories. The "historical digressions" are short and interesting as are the many exercises.

R.H. Battin, *An Introduction to the Mathematics and Methods of Astrodynamics*. American Institute of Aeronautics and Astronautics,

New York, 1987.

A major reference work from the basic two-body problem to highly sophisticated problems of engineering space dynamics with many fascinating personal remarks describing the U.S. space program of which the author was one of the major contributors. The large amount of material included and many mathematical details require devoted readership. Battin's earlier book, entitled *Astronautical Guidance,* McGraw Hill, Inc. (1964) is another volume of considerable importance to space engineering. The Apollo guidance system is discussed in detail by the author who was the director of the space guidance analysis division of this project.

A. Beer and K.A. Strand (editors), *Copernicus. Vistas in Astronomy.* Vol. 17, Pergamon Press, New York, 1975.

This volume represents the proceedings of a conference held in 1972 in Washington, D.C. to commemorate the 500th anniversary of the birth of Copernicus. It highlights Copernicus' ideas and place in the history of celestial mechanics.

G.D. Birkhoff, *Dynamical Systems.* American Mathematical Society, New York, 1927.

Advanced theoretical treatment emphasizing the qualitative aspect of dynamics. Main subjects are stability, periodic orbits, problem of three bodies and integrability.

D. Brouwer and G.M. Clemence, *Methods of Celestial Mechanics.* Academic Press, New York, 1961.

This advanced reference and textbook is one of the classics in the field of celestial mechanics. Astronomically oriented and dedicated readers will find a large amount of very useful information.

E.W. Brown, *An Introductory Treatise on the Lunar Theory.* University Press, Cambridge, 1896. Also Dover, New York, 1960.

A clear description of the problem of the motion of the Moon with several approaches to the solution. The major methods (Laplace, de Pontecoulant, Hansen, Delaunay, Hill, etc.) are explained and compared. This is a major reference book rather than an introductory treatment.

C.V.L. Charlier, *Die Mechanik des Himmels.* Vols. 1 and 2, Viet and Co., Leipzig, 1902-1907.

Mathematically oriented, highly advanced, but quite readable German text.

J. Chazy, *Mécanique Céleste*. Presses Universitaires de France, Paris, 1953.

Delightfully short and concise French text on a rather advanced level, discussing canonical transformations, variational equations, perturbation theories, etc.

G.A.   Chebotarev, *Analytical and Numerical Methods of Celestial Mechanics*. Nauka, Moscow, 1965. English translation by L. Oster, American Elsevier, New York, 1967.

Advanced, astronomically oriented, clearly written, graduate textbook. Planetary theories, lunar theories, study of minor planets, satellites and comets.

J.M.A. Danby, *Fundamentals of Celestial Mechanics*. Macmillan, New York, 1962.

The use of vectors can simplify some of the equations and ideas of celestial mechanics as demonstrated in this basically introductory book with several well-selected exercises.

R. Deutsch, *Orbital Dynamics of Space Vehicles*. Prentice-Hall, Englewood Cliffs, New Jersey, 1963.

Advanced text emphasizing the theoretical foundations and outlining the solution techniques used in celestial mechanics. The subjects treated in some detail, besides the basics, are orbit determination, analytical dynamics as applied to general perturbation techniques, and modern topological research related to the problem of three bodies.

G.N. Duboshin, *Celestial Mechanics, Fundamental Problems and Methods*. Nauka, Moscow, 1968 and *Celestial Mechanics, Analytic and Quantitative Methods*. Nauka, Moscow, 1978.

These basic and important books in our field are available only in their original Russian editions.

D. Dubyago, *The Determination of Orbits*. Nauka, Moscow, 1949. English translation by R.D. Burke, G. Gordon, L.N. Rowell and F.T. Smith, Macmillan, New York, 1961.

Book is basically for astronomers, discussing the determination of orbits of minor planets, comets and meteors. Many excellent examples.

K.A. Ehricke, *Space Flight*. Vol. I, Environment and Celestial Mechanics, 1960. Vol. II, Dynamics, 1962. Vol. III, Operations, 1964. D. Van Nostrand, Princeton, New Jersey.

Space missions and system analysis are emphasized in these easy to read volumes, directed to astronautical engineers, with details on ballistics and powered flights. Many well selected examples and references.

B. Erdi, *Egi Mechanika (Celestial Mechanics)*. Vols. 1-3, Eötvös University, Budapest, 1972-74.

Subjects emphasized are orbit determination, general perturbations, lunar theory, dynamics of artificial satellites. This clearly formulated basic text is available only in Hungarian.

P.R. Escobal, *Methods of Orbit Determination*. R.E. Krieger, Huntington, New York, 1965.

Oriented to aerospace engineers and applied astrodynamicists, this highly readable book offers a large amount of very useful information with many examples and detailed computational algorithms.

E. Finlay-Freundlich, *Celestial Mechanics*. Pergamon Press, New York, 1958.

Delightfully short and clear. The problem of two bodies is treated in the introductory chapter which is followed by advanced dynamical astronomy.

P.M. Fitzpatrick, *Principles of Celestial Mechanics*. Academic Press, New York, 1970.

Emphasis on artificial satellites, including rigid body rotational motion. Advanced and often mathematically oriented treatment with many exercises.

K.F. Gauss, *Theoria Motus Corporum Coelestium in Sectionibus Conicis Solem Ambientium (Theory of the Motion of the Heavenly Bodies Revolving around the Sun in Conic Sections)*. Hamburg, 1809. English translation by C.H. Davis, Little, Brown, Boston, 1957.

One of the classics of theoretical and computational celestial mechanics, emphasizing orbit determination (such as orbits from three observations), method of least squares, orbits of Ceres, Pallas and Juno, etc.

H. Goldstein, *Classical Mechanics*. Addison-Wesley, Reading, Mass., 1950.

Introductory (compared to Whittaker's book), easy to read, clear, basic reference text.

Y. Hagihara, *Celestial Mechanics*. Vols. 1 and 2, Massachusetts Institute of Technology Press, Cambridge, Mass., 1970-1972; Vols. 3-5, Japan Society for Promotion of Science, Tokyo, Japan, 1974-1976.

These volumes represent the encyclopedia of celestial mechanics. The reader will find everything discussed in considerable detail, even those subjects or approaches which are only indirectly connected with celestial mechanics. This major reference book is comprehensive, clear and connects astronomy with mathematics.

S.W. Hawking and W. Israel, *300 Years of Gravity*. Cambridge University Press, London, 1987.

Several contributors discussing mostly physics since Newton: cosmology, relativity, black holes, superstring unification and quantum theory.

P. Herget, *The Computation of Orbits*. Edwards Brothers, Ann Arbor, Michigan, 1948.

This astronomically oriented, easy to read, advanced text specializes in orbit determination. The author's and his associates' humor will be enjoyed by those scholars who carefully translate the Greek language subtitles of the chapters. For instance: Chapter 2. Problems in Spherical Astronomy. (Leave hope behind all ye who enter here.) Chapter 6. Improvement of the Orbit. (If at first you do not succeed try, try again.) Chapter 7. Special Perturbations. (These numbers laid end to end would reach to insanity.)

S. Herrick, *Astrodynamics*. Vols. 1 and 2, Van Nostrand Reinhold, London, 1971-1972.

Basic text with applications to space engineering and astronomy. Some of the unconventional notations require careful attention. Principle subjects are special and general perturbation theories, orbit determination and universal variables. Many examples with details.

D.G. King-Hele, *Satellites and Scientific Research*. Routledge and Kegan Paul, London, 1962 and *Theory of Satellite Orbits in an Atmosphere*. Butterworths, London, 1964.

One of the important contributors to the still unsolved problem of atmospheric effects on satellite motion. See also his Technical Memorandum No. 212 of the Royal Aircraft Establishment, entitled "A View of Earth and Air," 1974.

A. Koestler, *The Sleepwalkers*. Macmillan, New York, 1959.

History of science from about 3000 B.C. to Newton. Book emphasizes history rather than science, and the "cold war" between humanities and sciences.

J. Kovalevsky, *Introduction a la Méchanique Céleste*. Librairie Armand Colin, Paris, 1963. English translation, D. Reidel, Dordrecht, Holland, 1967.

> This short and well-written book emphasizes perturbation theories and the problem of artificial satellites. In the author's opinion, "the construction of increasingly more accurate analytical theories remains the central task of celestial mechanics."

J. Lagrange, *Oeuvres*. 14 Volumes, Gauthier-Villars, Paris, 1867-1892 and *Méchanique Analytique*. Veuve Desaint, Paris, 1788.

> Author's infatuation with equations makes his work somewhat difficult to read. The brilliance of the presentations and the analytical manipulations show clearly Lagrange's mastery of his subjects and will fill the reader with admiration of these volumes written in French.

P.S. Laplace, *Traité de Mécanique Céleste*. Vols. 1-3, Duprat, Paris, 1799-1802; Vol. 4, Courrier, Paris, 1805; Vol. 5, Bachelier, Paris, 1823-1825. English translation by N. Bowditch, Volumes 1-4, Chelsea, New York, 1829-1839.

> This major classic of celestial mechanics is directed to advanced, astronomically oriented readership. The comments and detailed explanations in the English translation are of considerable help.

A.B. Lerner, *Einstein and Newton*. Lerner, Minneapolis, Minnesota, 1873.

> Lives, backgrounds and accomplishments compared via many valuable quotations from correspondences.

S.W. McCuskey, *Introduction to Celestial Mechanics*. Addison-Wesley, Reading, Massachusetts, 1963.

> This easy to read, clear, introductory text emphasizes the basic principles and offers well selected examples.

J. Moser, *Stable and Random Motions in Dynamical Systems with Special Emphasis on Celestial Mechanics*. Princeton University Press, Princeton, N.J., 1973.

> The author of this mathematical exposition is the third member of the famous KAM trio (Kolmogorov, Arnold, Moser). The book concentrates on the convergence problem of the series solutions, the role of small divisors, non-integrability, ergodic motions, quasi-periodic behavior, etc. With patience and some analytical background the reader will enjoy this version of Moser's Hermann Weil lecture series delivered at the Institute for Advanced Study in 1972.

F.R. Moulton, *An Introduction to Celestial Mechanics*. Macmillan, New York, 1960.

   This book is an easy to read classic with astronomical orientation. The historical remarks and well-selected references are most informative.

I.   Newton, *Philosophiae Naturalis Principia Mathematica* (*The Mathematical Principles of Natural Philosophy*). Royal Society of London, 1687. Translation by A. Motte (1729) edited by F. Cajori, University of California Press, Berkeley, California, 1946.

   Difficult to read because of the author's well known insistence of using classical geometry to treat dynamical problems. It is reasonable to assume that Newton derived many of his theorems by calculus but presented the proofs with geometrical tools. To quote Laplace: "The *Principia* is pre-eminent above any other production of human genius." The First Book is on theoretical mechanics, containing his laws of motion, law of gravitation, and problem of two bodies. The Second Book is on hydrodynamics. The Third Book forms the major part of the *Principia* and treats the dynamics of satellites and planets, including the problem of mass determination, flattening, and precession of the equinoxes. Lunar theory, theory of tides and orbits of comets conclude the Third Book.

L.A. Pars, *Treatise on Analytical Dynamics*. Heinemann, London, 1965.

   Excellent presentation of dynamics, emphasizing the principle of least action. Many references and examples.

H.C. Plummer, *An Introductory Treatise on Dynamical Astronomy*. Cambridge University Press, London, 1918. Reprinted by Dover, New York, 1960.

   This book is strongly oriented to dynamical astronomy in a clear and understandable style, on an advanced level. In addition to the basic problems of two and three bodies, considerable details are offered on planetary theory, lunar theory, determination of orbits, orbits of binary stars, perturbation methods, etc.

H. Poincaré, *Les Méthodes Nouvelles da la Mécanique Céleste*. Gauthier-Villars, Paris, 1892-1899. Reprinted by Dover, New York, 1957. English translation NASA-TTF-450, 1967.

   It is difficult to read because of the author's concise style, but it has many brilliant ideas and suggestions for future research in celestial mechanics, especially along mathematical lines.

H. Pollard, *Mathematical Introduction to Celestial Mechanics*. Prentice-Hall, Englewood Cliffs, New Jersey, 1966.

> The author shows how a mathematician might write an easy to read and clear book. Some of the subjects emphasized in the book might be of more interest to mathematicians than to astronomers or to engineers.

I. Prigogine, *From Being to Becoming*. W.H. Freeman, San Francisco, 1980.

> Book connects the "classical period" of dynamics with the modern aspects of statistical mechanics. The reader will find a refreshing view of instability, non-integrability, and the complexities which uncertainties can create in the physical sciences.

A.E. Roy, *Orbital Motion*. Adam Hilger, Bristol, 1978.

> This book is an easy to read, excellent text, offering combinations of astronomical and space research applications. Many advanced subjects are covered and several useful exercises are included.

Y. Ryabov, *Celestial Mechanics*. Foreign Languages Publishing House, Moscow, 1959. Also Dover, New York, 1961.

> This semi-popular book is simple and presents basic ideas without much mathematics and with many numerical explanations. For the layperson the book is heavy reading, but the undergraduate astronomy or engineering student will consume it with pleasure in a short time with considerable benefit.

C.L. Siegel and J.K. Moser, *Lectures on Celestial Mechanics*. Springer-Verlag, Berlin, 1971.

> This is a readable, mathematically oriented book, based on Siegel's book published in 1956. It concentrates on the problem of solution of differential equations of celestial mechanics and on the problem of three bodies. It is based on the works of Poincaré, Sundman, Birkhoff and Wintner.

W.M. Smart, *Celestial Mechanics*. Longmans, Green, London, 1953.

> One of the few easy to read, advanced books on celestial mechanics, with astronomical orientation. Almost one third of the book deals with lunar theories. Few references and no exercises are given but complete and clear explanations of the basic theoretical aspects are offered.

T.E. Sterne, *An Introduction to Celestial Mechanics*. Interscience Publishers, New York, 1960.

This short and clear text is for advanced level readers. It contains the author's original contributions to the analysis of the dynamics of artificial satellites.

E.L. Stiefel and G. Scheifele, *Linear and Regular Celestial Mechanics*. Springer-Verlag, Berlin, 1971.

Book describes regularizing transformations which result in linear differential equations for the problem of two bodies. Generalizations are discussed from two to three dimensions, using canonical theories. Several results are shown, concerning the numerical advantages of regularization.

K. Stumpff, *Himmelsmechanik*. Vols. 1-3, Deutscher Verlag der Wissenschaften, Berlin, 1959-1974.

Clear, easy to read treatment emphasizing orbit determination, the problem of three bodies and perturbation theories. Many details make reading this book, written in German, a rather lengthy but definitely worthwhile project.

V. Szebehely, *Theory of Orbits*. Academic Press, New York, 1967.

This advanced reference text discusses the restricted problem of three bodies with theories and applications concerning periodic orbits, space trajectories, stability and dynamical astronomy.

L.G. Taff, *Celestial Mechanics: A Computational Guide for the Practitioner*. J. Wiley, New York, 1985.

This highly readable, often informal text is prepared for advanced courses, emphasizing orbit determination and computational approaches.

F. Tisserand, *Traité de Mécanique Céleste*. Vols. 1-4, Gauthier-Villars, Paris, 1889-1896.

This book, written in French, is a remarkable advanced text directed to astronomically oriented readers. It is more readable than some of the other classics, such as Poincaré or Laplace. Contains discussions of general perturbation methods, figures of celestial bodies and their rotational motions, theory of the motion of the Moon, theory of motion of Jupiter's and Saturn's satellites and of minor planets and comets.

W.T. Thomson, *Introduction to Space Dynamics*. J. Wiley, New York, 1961.

More along the lines of gyrodynamics, optimization, flexible missiles, etc. than celestial mechanics. Many examples of practical importance.

E.T. Whittaker, *A Treatise on the Analytical Dynamics of Particles and Rigid Bodies*. Cambridge University Press, London, 1904.

This is one of the excellent books from which one can learn advanced dynamics. Many applications to celestial mechanics are treated in a clear style.

A. Wintner, *The Analytical Foundations of Celestial Mechanics*. Princeton University Press, New Jersey, 1941.

The author wishes to emphasize "analytical foundations" in the title pointing out that "it is almost forgotten how much the theory of analytic functions owes to the elementary problem of two bodies." This book was written by and is appreciated by "ε-trained mathematicians" more than by engineers and astronomers. Excellent set of references are given, and it is carefully pointed out that the traditional references to the origin of the fundamental mathematical notions in analytical dynamics are often incorrect.

Subject Index

Name Index

Phoenix Bookstore
Lambertville, NJ
Sat 23 Jan 1999 ~ #13